Stephen Hawking: Genius at Work

Stephen Hawking: Genius at Work

Explore his life, mind and science through the objects in his study

Roger Highfield

SCIENCE MUSEUM

Contents

Foreword

The image of Stephen Hawking in his motorized wheelchair, with head slightly to one side and hands crossed over the controls, caught the public imagination, as a symbol of the triumph of mind over matter. As with the Delphic oracle of ancient Greece, physical impairment seemed compensated by almost supernatural gifts, which allowed his mind to roam the universe freely, upon occasion enigmatically revealing some of its secrets.

Of course, such a romanticized image can represent but a partial truth. Those who knew Stephen would clearly appreciate the dominating presence of a real human being, with an enormous zest for life, great humour, and tremendous determination, yet with normal human weaknesses.

Despite his terrible physical circumstance, he almost always remained positive about life. He enjoyed his work, the company of other scientists, the arts, the fruits of his fame, his travels. He took great pleasure in children, sometimes entertaining them by swivelling around in his motorized wheelchair. Social issues concerned him. He promoted scientific understanding. He could be generous and was very often witty. On occasion he could display something of the arrogance that is not uncommon among physicists working at the cutting edge, and he had an autocratic streak. Yet he could also show a true humility that is the mark of greatness.

This book opens an unusual window on many aspects of his remarkable life by discussing the objects in his Cambridge office that were acquired by the nation for the Science Museum Group. From the books and papers to his awards and pictures, every item has a story to tell, but here I would like to focus on the narrative that is perhaps the most difficult to describe through objects alone. I should like to indicate his greatly impressive, sometimes revolutionary, contributions to physics, and why Stephen Hawking was extremely highly regarded by his scientific peers.

Hawking had been diagnosed shortly after his twenty-first birthday as suffering from an unspecified incurable disease, which was then identified as motor neurone disease amyotrophic lateral sclerosis, or ALS. Soon afterwards, rather than succumbing to depression, as others might have done, he began to set his sights on some of the most fundamental questions concerning the physical nature of the universe.

In 1964, when Hawking was in his second year at Cambridge, I (then at Birkbeck College in London) had established a mathematical theorem which showed, on the basis of a few assumptions, that a collapsing over-massive star would result in a singularity in space-time – a place where it would be expected that densities would become infinite – what we now refer to as a "black hole". There, Einstein's classical theory of gravity, general relativity, would have reached its limits.

Hawking had also been thinking about this kind of problem and his supervisor Dennis Sciama made a point of bringing us together. It did not take Hawking long to find a way to use my theorem in an unexpected way, so that it could be applied (in a time-reversed form) to show that a big-bang-type singularity was a necessary implication of Einstein's general relativity.

A few years later, in 1970, he and I joined forces to publish an even more powerful theorem which showed how the black holes that we expect to find in nature conform to a specific geometry. Following this work, Stephen established some remarkable analogies between the behaviour of black holes and the basic laws of thermodynamics, the science of

heat and work. His research in classical general relativity was the best anywhere in the world at that time.

Hawking then turned his attention to quantum effects in relation to black holes and was startled to find that they would radiate away their energy – this radiation, that Hawking predicted, coming from black holes, is now, very appropriately, referred to as Hawking radiation. These achievements are generally regarded as Hawking's greatest contributions. That they have deep implications for future theories of fundamental physics is undeniable, though the detailed nature of these implications is still a matter of much heated debate.

Following his black-hole work, and until the end of his life, Hawking turned his attentions to the problem of quantum gravity, which involves correctly imposing the quantum procedures of particle physics on to the structure of space-time described by general relativity, which is generally regarded as the most fundamental unsolved foundational issue in physics.

He became increasingly involved with the popularization of science. This began with *A Brief History of Time* in 1988.

In the wake of the book's astounding success, he took great delight in his commonly perceived role as "the No 1 celebrity scientist"; huge audiences would attend his public lectures, perhaps not always just for scientific edification.

In 2020, two years after Stephen died, I was jointly awarded the Nobel Prize, along with two other physicists, for my 1964 research on relativity and black holes. I am often asked if I thought that, had he survived, Stephen could have shared the prize too. It is not clear to me that the Nobel committee would have made this decision because it appears to be a requirement that such theoretical work needs observational confirmation and this is the kind of thing that my colleagues who were co-awarded the prize actually recently achieved. There is, however, a definite possibility that experimental evidence of Hawking radiation might, in time, emerge, which is the sort of thing which would be certainly worthy of a Nobel Prize.

Sir Roger Penrose

Sir Roger Penrose, physicist, mathematician and philosopher of science, is the Emeritus Rouse Ball Professor of Mathematics at the Mathematical Institute of the University of Oxford. This is an abridged, edited and updated version of an article he wrote for *The Guardian* newspaper in 2018.

Preface
My Brief History of Hawking

I fell into Stephen Hawking's orbit more than three decades ago during a trip to California. It was March 1988 and our paths crossed in Berkeley, where he made a huge impression, not just on me but everyone he encountered.

He was promoting *A Brief History of Time*, his runaway international bestseller which had made him the most famous scientist on the planet, while I was on a tour of countless research institutes across America in my role as the Technology Correspondent of the *Daily Telegraph*. A public affairs officer for the University of California, Berkeley, had urged me to come along to hear Hawking, adding that he'd never witnessed such feverish anticipation for a talk by a visiting scientist.

The university's main auditorium and satellite venues were fit to burst as the public gathered to see Stephen Hawking, hear his synthetic accent (beginning his talk, as ever, with his trademark phrase, "Can you hear me?") and learn what he thought about cosmology and astrophysics. He was given a rock-star reception.

But while I was impressed by the hoopla, I was also moved by the number of people in wheelchairs, who seemed to draw inspiration from being close to him and, presumably, took heart from his extraordinary story: even though he had been given a death sentence 25 years earlier, with a devastating diagnosis of motor neurone disease, he had defied the odds to live on for decades to glimpse the deepest secrets of the Big Bang, black holes and the universe.

Hawking and I were staying at the same hotel in Berkeley, and I was also struck by how, rather than remain in his room to be fed by his carers, Hawking and his entourage would come down to the restaurant. Despite his pitiless ailment and difficult circumstances, about which he never grumbled, he had a remarkable zest for life and adored being in the public eye.

OPPOSITE Roger Highfield's *Daily Telegraph* article on Hawking's synthetic voice, 1988.

LEFT Roger Highfield (far left), just before he encouraged visitors to sing happy birthday to Stephen Hawking at an exhibit in the Science Museum to mark Hawking's seventieth birthday.

Picture: MICHAEL WEBB

Prof Stephen Hawking at work: 'My computer system is vital ... Without it I would be a vegetable, unable to communicate'

Finding a voice with a flick of the finger

TECHNOLOGY

Roger Highfield

ONE OF THE most brilliant theoretical physicists of our time — Cambridge University's Prof Stephen Hawking — is a striking example of how, with the help of computers, the human intellect can triumph over disability.

He is now widely considered one of the greatest theoretical physicists since Einstein. Yet, at 46, Hawking has lived more than 20 years longer than he once expected to.

He is one of the disabled people who last Wednesday at London's Dorchester Hotel helped show representatives of the Government, employers, trade unions and computer companies how disabled people have become eminent in their own fields with the help of computers.

Deprived of virtually all voluntary movement by years of wasting illness, Prof Hawking guides his battery-powered wheelchair — and operates its built-in computer and voice synthesiser — with a barely perceptible movement of a finger.

He says he is "all in favour" of the initiative to use high technology to help more disabled people that has been launched by the British Computer Society. "A large number of disabled people could be helped by modern information technology," he says.

He speaks with a synthesised American accent. "My computer system is vital to me. Without it I would be a vegetable, unable to communicate."

However, he feels that not enough attention is being paid in this country to easing the lot of the disabled. "There is

much less done in Britain for disabled people than in other countries, such as the United States or Sweden."

When he talks he squeezes a slim black control box, his head lolling back against a head-rest. A series of clicks issues from the computer. A personal dictionary of about 3,000 words scrolls down a screen, and Prof Hawking is able to move his finger a fraction to squeeze the control box and halt a cursor on the required word.

When his answer is complete, the voice synthesiser, made by the Californian company Speech Plus, speaks. The voice "is the best I have heard, though it gives me an accent that has been described variously as Scandinavian, American or Scottish".

He was recently offered an upgraded version of the voice but, says Mr Walt Waltosz, president of Words Plus, the company that developed the software, "he didn't like it. He said that is not my voice and

made them put the old one back." Mr Woltosz says the key to setting up a system for a disabled person is to find a way for them to communicate with the computer. Some can use a keyboard, others, like Prof Hawking, use a switch.

Infra-red switches, tiny joysticks, blinking, eye movement or eye gaze, are also used. A "morse code" system has been developed for blind disabled people.

"There is something available for almost any disabled person," Mr Waltosz says.

Using his hand switch, Prof Hawking is able to write lectures and save them on disc. "I can then send it to the speech synthesiser a sentence at a time," he says.

Recently he gave a series of lectures at the University of California at Berkeley, a campus that is striking because of the large number of disabled people that can be seen whizzing around on wheelchairs.

His lectures were packed out — professors and students charged into the university's largest lecture theatre to hear him talk about his goal: a complete understanding of the universe.

To this end, he has been working since 1974 on marrying the two cornerstones of 20th-century physics: Einstein's General Theory of Relativity, which deals with gravity and the study of large scale phenomena, and quantum theory, which covers elementary particles and the study of the very small.

He has proposed a model of the universe based on two concepts of time: "real time", or time as human beings experience it from their limited standpoint; and "imaginary time", the time on which the world might really run.

Prof Hawking studied physics at Oxford University, and when his illness first struck, he was a first-year research student at Cambridge. He was diagnosed in his twenties as having amyotrophic lateral sclerosis (ALS), a paralysing disease that attacks the body's nervous system.

"The only consolation they could give me was that I was not a typical case. There did not seem much point in working at my research, because I didn't expect to live long enough to finish my PhD," Prof Hawking says.

But the spread of the disease slowed, and he met Jane, his future wife. They now have three children. "This gave me something to live for," Hawking says. "If we were to get married, I had to get a job. And to get a job, I had to finish my PhD. I therefore started working hard for the first time in my life. To my surprise, I found I liked it."

The images, items, books, papers and bric-a-brac [in his office] offer an unusual perspective on both Stephen Hawking himself and his inner circle.

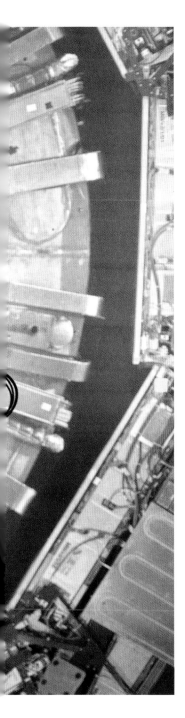

ABOVE Stephen Hawking and Peter Higgs visiting the Science Museum, London, in 2013 to see the Collider exhibition.

My report on his talk ("A universal way with words") ended up on the front page of the *Daily Telegraph*: "he held a packed audience spellbound during a lecture yesterday," I wrote, "without saying a word". Hawking told us how, during its creation, the universe borrowed heavily from its gravitational energy to finance the creation of more matter. "The result was a triumph for Reagan economics," he said, in reference to the policies of the then President, Ronald Reagan.

It was the first of many stories I wrote about Hawking for the newspaper, mostly as the *Telegraph*'s Science Editor, including a story that appeared in September 2008 on whether the Large Hadron Collider, the biggest and most complex particle accelerator, would seed a black hole to swallow the Earth. Reassuringly for our readers, Stephen had predicted that "the world will not come to an end", explaining how the energies in the LHC were feeble compared with events in the rest of the universe. "If a disaster was going to happen, it would have happened already."

And, indeed, we all lived to see the Collider find evidence of the Higgs particle – which, roughly speaking, gives all other particles their mass – despite a bet laid by Stephen Hawking against its discovery. He had mischievously joked at a Science Museum event that physics would have been "far more interesting" if we had not been able to find the Higgs.

Any account of Stephen Hawking's life is subjective and partial, and depends on his dealings with the outside world, whether mediated by his public statements, family, university publishers, papers or, of course, how these sources were refracted through the lens of the international media.

That is why I was so excited by the opportunity to write about his office for the Science Museum, after its contents were acquired by the nation. The images, items, books, papers and bric-a-brac offer an unusual perspective on both Stephen Hawking himself and his inner circle, providing a relatively unfiltered look at Hawking the brilliant scientist, communicator, activist and global icon.

In these objects are reflected his life, obsessions and science, even down to the most mundane items, because they were essential to the way he lived or critical for the way he worked, or because he wanted them around him to summon up happy memories, celebrate triumphs or record his passions and interests.

The contents of his office joined the Science Museum Group's Collection in 2021, and since then a dedicated team of colleagues has had the joy of sorting through, cataloguing, preserving and researching them, bringing new discoveries and connections to light, many of which are presented in the following pages.

Is it intrusive to pore over these contents? No. The office was a space that Hawking himself had curated, with visitors very much in mind. Indeed, remarked Juan-Andres Leon, my colleague and the curator of physics in charge of the contents, the office is, fittingly, a microcosm of the late twentieth century and Stephen Hawking's influence on it. "I never met Stephen in person," he said, "but for people of my generation he was a presence throughout my life. As a child I aspired to one day read and understand that book with the blue spine on everyone's bookshelves: *A Brief History of Time*.

People like me grew up with Stephen, sometimes first as a character on TV, but later, while I was studying 'grown-up' physics, I could see his influence on today's understanding of the universe."

His office is a time machine, and not just into the past. Stephen grappled daily with difficult questions about our future and how to live, as our bodies reach their limits and as more of our existence spills into the electronic and digital realms. His office and its contents will be a significant resource for historians and researchers for generations to come, and an enduring fascination for anyone interested in one of the best-known scientists ever.

I was not an acquaintance or a colleague of Stephen Hawking, let alone a friend. Over the decades my job meant keeping abreast of science, and that meant keeping up with what he was doing, saying and thinking. However, over the years our lives intersected many times as I covered his science, and his story, first for the *Daily Telegraph*, then as Editor of *New Scientist*, and now as the Science Director of the Science Museum Group. I found myself inside Hawking's Cambridge office at various times down the years, gazing around at the remarkable paraphernalia, photographs, books, bets, pre-prints and papers.

Over the years, the office has morphed and changed. Gone is the jokey sign that read: "Yes, I AM The Centre Of The Universe", along with the swirling cloud of vapour on his desk sent up by his humidifier, a present from his second wife, Elaine, to ease his breathing, which she had created from a vaporizer and laminated bowl, painted gold inside, and decorated with seashells and stones. Gone also are the photographs of his children, grandchildren and family, which were not part of the museum's acquisition. Yet other features of the office remained constant for decades, such as a baffling but intriguing blackboard crammed with pun-laden in-jokes and scientific "graffiti".

Most memories are vivid, a handful embarrassingly so. I can remember when, as Science Editor of the *Daily Telegraph*, I took a call from Hawking's personal assistant, putting him on the line to discuss an article I had asked him to write. I tried hard to focus on the synthesized voice at the other end of the line but could not make out much of what he was saying against the thrum of the *Telegraph*'s bustling, open-plan office. I was too embarrassed to ask him to repeat his pitch again after his second attempt. That sounded brilliant, I assured him, having absolutely no idea what I had commissioned him to write. And it was.

Another toe-curling discovery was that he could render me star-struck. In November 2001, I did an event with Hawking at the Institute of Education in London, fielding him questions from a large audience of *Telegraph* readers to mark the publication of *The Universe in a Nutshell*, yet another of his books that expanded the cosmic horizons of millions of readers. I found myself gabbling away, filling the long gaps between his witty responses with increasingly inane patter. I learned the hard way that it was well worth the wait for his sparkling one-liners – he wouldn't compare the joy of discovery to sex, he once joked, but it does last longer. Like his humour, his concise quips were a product of necessity, given the constraints of using his voice synthesizer.

I can only assume he was used to being quizzed by a slightly flustered interviewer: afterwards, in the green room, I was given a "signed" copy by his then wife, Elaine,

OPPOSITE *Daily Telegraph* serialization of Hawking's popular book *The Universe in a Nutshell*, published in 2001.

20   • • •   Wednesday, October 17, 2001   THE DAILY TELEGRAPH

Science

The future of mankind... in a

By Stephen Hawking

In the first of three exclusive extracts from *The Universe in a Nutshell*, Prof Stephen Hawking's long awaited sequel to *A Brief History of Time*, the world's best known cosmologist discusses his hopes and fears for the future, from GM humans to alien life

The reason *Star Trek* is so popular is because it is a safe and comforting vision of the future. I'm a bit of a *Star Trek* fan myself, so I was easily persuaded to take part in an episode in which I played poker with Newton, Einstein, and Commander Data. I beat them all, but unfortunately there was a red alert, so I never collected my winnings.

Star Trek shows a society that is far in advance of ours in science, in technology, and in political organisation. (The last might not be difficult.) There must have been great changes, with their accompanying tensions and upsets, in the time between now and then, but in the period we are shown, science, technology, and the organisation of society are supposed to have achieved a level of near perfection.

I want to question this picture and ask if we will ever reach a final steady state in science and technology. At no time in the 10,000 years or so since the last ice age has the human race been in a state of constant knowledge and fixed technology.

There have been a few setbacks, like the Dark Ages after the fall of the Roman Empire. But the world's population, which is a measure of our technological ability to preserve life and feed ourselves, has risen steadily, with only a few hiccups such as the Black Death.

In the last 200 years, its growth has become exponential; that is, the population grows by the same percentage each year. Currently, the rate is about 1·9 per cent a year. That may not sound like very much, but it means that the world population doubles every 40 years.

Other measures of technological development in recent times are electricity consumption and the number of scientific articles. They, too, show exponential growth, with doubling times of less than 40 years. There is no sign

that scientific and technological development will slow down and stop in the near future – certainly not by the time of *Star Trek*, which is supposed to be not that far in the future. But if the population growth and the increase in the consumption of electricity continue at their current rates, by 2600 the world's population would be standing shoulder to shoulder, and electricity use would make the Earth glow red hot.

If you stacked all the new books being published next to each other, you would have to move at 90 miles an hour just to keep up with the end of the line. Of course, by 2600 new artistic and scientific work will come in electronic forms, rather than as physical books and papers. Nevertheless, if the exponential growth continued, there would be 10 papers a second in my kind of theoretical physics, and no time to read them.

Clearly, the present exponential growth cannot continue indefinitely. So what will happen? One possibility is that we will wipe ourselves out completely by some disaster, such as a nuclear war. There is a sick joke that the reason we have not been contacted by extra-terrestrials is that when a civilisation reaches our stage of development, it becomes unstable and destroys itself. However, I'm an optimist. I don't believe the human race has come so far just to snuff itself out when things are getting interesting.

The *Star Trek* vision of the future – that we achieve an advanced but essentially static level – may come true in respect of our knowledge of the basic laws that govern the universe. As I shall describe in a forthcoming extract in *The Daily Telegraph*, there may be an ultimate theory that we will discover in the not-too-distant future. This ultimate theory, if it exists, will determine whether the *Star Trek* dream of warp drive can be realised. According to

present ideas, we shall have to explore the galaxy in a slow and tedious manner, using spaceships travelling slower than light; but since we don't yet have a complete unified theory, we can't quite rule out a *Star Trek* style warp drive.

On the other hand, we already know the laws that hold in all but the most extreme situations: the laws that govern the crew of the Enterprise, if not the spaceship itself. Yet it doesn't seem that we will ever reach a steady state in the uses we make of these laws or in the complexity of the systems that we can produce with them.

By far the most complex

systems that we have are our own bodies. Life seems to have originated in the primordial oceans that covered the Earth four billion years ago. How this happened we don't know. It may be that random collisions between atoms built up large molecules that could reproduce themselves and assemble themselves into more complicated structures. What we do know is that by three and a half billion years ago, the highly complicated molecule DNA had emerged.

DNA is the basis for all life on Earth. It has a double helix structure, like a spiral staircase, which was discovered by

Francis Crick and James Watson in the Cavendish lab at Cambridge in 1953. The two strands of the double helix are linked by pairs of nucleic acids, like the treads in a spiral staircase. There are four kinds of nucleic acids: cytosine, guanine, tyrosine and adenine.

The order in which the different nucleic acids occur along the spiral staircase carries the genetic information that enables the DNA molecule to assemble an organism around it and reproduce itself. As the DNA makes copies of itself, there are occasional errors in the order of the nucleic acids along the

spiral. In most cases, the mistakes in copying make the DNA either unable or less likely to reproduce itself, meaning that such genetic errors, or mutations, as they are called, will die out.

But in a few cases, the error or mutation will increase the chances of the DNA surviving and reproducing. Such changes in the genetic code will be favoured. This is how the information contained in the sequence of nucleic acids gradually evolves and increases in complexity.

Because biological evolution is basically a random walk in the space of all genetic possibilities, it has been very

Above: from Big Bang to modern man – the human race has been in existence for only a tiny fraction of the history of the universe. Any alien life we meet is likely to be much more primitive or much more advanced than we are

slow. The complexity, or number of bits of information, that is coded in DNA is roughly the number of nucleic acids in the molecule. For the first two billion years or so, the rate of increase in complexity must have been of the order of one bit of information every 100 years.

The rate of increase of DNA complexity gradually rose to about one bit a year over the last few million years. But then, about six or eight thousand years ago, a major new development occurred. We developed written language. This meant that information could be passed from one generation to the next without

having to wait for the very slow process of random mutations and natural selection to code it into the DNA sequence. The amount of complexity increased enormously. A single paperback romance could hold as much information as the difference in DNA between apes and humans, and a 30-volume encyclopedia could describe the entire sequence of human DNA.

Even more important, the information in books can be updated rapidly. The current rate at which human DNA is being updated by biological evolution is about one bit a year. But there are 200,000 new books published each year, a new-information rate of over a million bits a second. Of course, most of this information is garbage, but even if only one bit in a million is useful, that is still 100,000 times faster than biological evolution.

This transmission of data through external, non-biological means has led the human race to dominate the world and to have an exponentially increasing population. But now we are at the beginning of a new era, in which we will be able to increase the complexity of our internal record, the DNA, without having to wait for the slow process of biological evolution.

There has been no significant change in human DNA in the last 10,000 years, but it is likely that we will be able to completely redesign it in the next thousand. Of course, many people will say that genetic engineering of humans should be banned, but it is doubtful we will be able to prevent it. Genetic engineering of plants and animals will be allowed for economic reasons, and someone is bound to try it on humans. Unless we have a totalitarian world order, someone somewhere will design improved humans.

Clearly, creating improved humans will create great social and political problems

Right: unless we have a totalitarian world order, someone somewhere will design improved humans. This may be necessary for long duration space travel or to outsmart machines

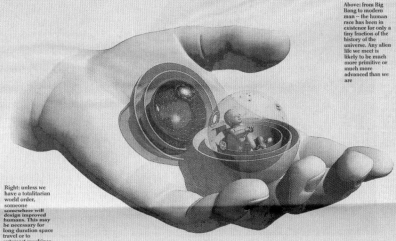

That sounded brilliant, I assured him, having absolutely no idea what I had commissioned him to write. And it was.

Roger Highfield —

The Universe in a Nutshell

Stephen Hawking

"Onward to the Brave New World"

Best wishes —
Stephen Hawking

BANTAM PRESS

LONDON · NEW YORK · TORONTO · SYDNEY · AUCKLAND

8th Nov. 2001.

OPPOSITE Roger Highfield's copy of *The Universe in a Nutshell*, signed by Hawking's second wife, Elaine, with his thumbprint.

inscribed, "Onward to the Brane New World" (branes featured in his latest quest for a theory of everything). She thanked me on his behalf and, in one swift motion, took his hand, rolled his thumb on an inkpad, and left his print on my copy.

At the Science Museum, we celebrated Stephen Hawking's science with exhibits and events. Even before I joined the Science Museum Group in 2011, I had started working on his seventieth birthday exhibit. As part of this celebration in the museum I helped to bring Hawking together with David Hockney for an iPad portrait, and the following years would see our paths cross again and again.

One of the most memorable encounters came a little later, in the wake of his seventieth birthday, when ill health had forced him to pull out of his celebrations in Cambridge and the VIP opening of the exhibit about him in the museum. Given his frequent visits to Papworth Hospital in Cambridge, we all feared the worst. But the following month, with just a day's notice, there he was in the museum to see the exhibit, including the specially commissioned David Hockney iPad portrait. It seemed somehow fitting that when Stephen Hawking arrived, he came with an entourage worthy of a social media sensation: there were assistants, carers, a green room and, of course, he had specific dietary demands too.

The draw of his fame was irresistible. In the Exploring Space Gallery of the Museum, an enormous crowd, including my wife and children, craned their necks to catch a glimpse of the genius who had explored the furthest reaches of the universe with his mind. The opportunity to catch up on his lost birthday celebrations seemed too good to miss: I asked the crowd to join me in singing a long overdue "Happy Birthday". He responded with a huge, spontaneous grin, then demanded to be taken on a tour of "one of my favourite places". He stayed all day.

Later that year, after the museum awarded him a fellowship in recognition of his academic work, Hawking spoke at the opening ceremony for the 2012 Paralympic Games in London, a thrilling spectacle watched by millions – including me – that combined soaring operatic performances, dance, punk and high-wire aerial stunts. Showing no sign of nerves, he shed a spotlight on the role of science in changing attitudes towards disability. "There is no such thing," he declared, "as a standard or run-of-the-mill human being." In his memoir, published the following year, he wrote that the ceremony had been "my biggest-ever audience".

Hawking returned to the museum the following year to celebrate the opening of our exhibition about the Large Hadron Collider, the world's most powerful particle accelerator, reiterating in his talk that the Science Museum was one of his favourite places. "I have been coming here for decades. And that simple fact, in itself, tells quite a story."

More details of his lifelong relationship with the museum emerged two years later, when he gave a personal tour of it to London's "Guest of Honour", a 24-year-old Californian schoolteacher, Adaeze Uyanwah , who had won an international competition organized by the official promotional organization for London. "When we were young," Hawking told her, "my mother used to leave me at the Science Museum, my sister Mary at the Natural History Museum, and my younger sister Philippa at the Victoria and Albert Museum. At the end of the day, my mother collected us all."

Among the Museum objects that he highlighted in his tour was Hockney's iPad portrait of him ("I'm still not quite sure about the fingers," he joked) and his speech synthesizer, which had been on loan to the museum since 1999. "I was happy to lend my voice recently to Eddie Redmayne to give him a bit of a boost in his efforts to win an Oscar," he told her, referring to the biopic *The Theory of Everything*. "Unfortunately, Eddie did not inherit my good looks."

Later that year, he attended yet another museum event, this time with the veteran Soviet cosmonaut, Alexei Leonov, who had survived the ordeal of the first spacewalk, and was the first space artist too. Leonov presented the Cambridge cosmologist with a sketch, and this veteran cosmonaut, spacefaring legend and hero of the Soviet Union twice over was delighted with Hawking's spontaneous response: "Stephen smiled – hooray!"

Despite his devastating diagnosis of motor neurone disease in 1963, when he was just 21 years old, Stephen Hawking only succumbed more than half a century later, on 14 March 2018, aged 76. Though a shock, his death was not unexpected, given how fragile his health had become. In the decades after that diagnosis his mind still managed to travel light years, helping to turn cosmology from a fringe subject into perhaps the most compelling of all the sciences, and one in which he provided profound insights into gravity, space and time that few have delivered since Albert Einstein, who had also achieved that rare feat among scientists of becoming a global celebrity. No doubt Stephen would have been amused that his death, which triggered tributes from Queen Elizabeth II to NASA, coincided with the day Einstein himself had been born.

Six months later, in October 2018, I found myself on the stage of the museum's IMAX theatre, as MC of an event to mark the publication of Stephen Hawking's final book, *Brief Answers to the Big Questions*. With me on the IMAX stage were two of his children, Lucy and Tim Hawking; the co-authors of his final research paper, Professor Malcolm Perry from the Department of Applied Mathematics and Theoretical Physics at the University of Cambridge, and Professor Andrew Strominger from Harvard University; and Professor Fay Dowker, a one-time student of Professor Hawking, now Professor of Theoretical Physics at Imperial College London. The press conference ended with an inspiring statement from Stephen Hawking himself, which had been recorded before he died. "People have always wanted answers to the big questions. *Where did we come from? How did the universe begin? Is there anyone out there?*"

When it comes to the big questions about Stephen Hawking's own life, passions and science, many answers lay in his office, and the extraordinary objects that it contained. Each one has a fascinating story to tell. In the following pages, you can glimpse the trappings of a singular scientific life, the workings of an extraordinary mind, along with the grit and determination that he needed to overcome huge physical challenges every day. I have yet to escape the extraordinary gravitational pull of his fame. I doubt I ever will.

OPPOSITE TOP Hawking gives London's "Guest of Honour" Adaeze Uyanwah a tour of the Science Museum in 2015.

OPPOSITE BOTTOM Cosmonaut Alexei Leonov and his sketch of Stephen Hawking at an event at the Science Museum, London, in 2015.

Hawking's Rosetta Stone

The contents of Stephen Hawking's office are the biographical equivalent of the Rosetta stone: collectively, the contents provide a key to unlock the secrets of his extraordinary science and no less extraordinary life. Behind this sign was the hub of the world's best-known physicist of recent years, a unique space around double the size of a regular university office that was tailored for his needs and working practices, and included a sink and kitchen area, along with a space for medical treatments. The offices of his technical and administrative assistants were right next door.

They were part of the University of Cambridge's Department of Applied Mathematics and Theoretical Physics, established in 1959 in Free School Lane. By the time Hawking joined the department his office was on a site between Silver Street and Mill Lane, but he ended up transferring with the department in 2002 to a new building in west Cambridge.

OVER THE THRESHOLD
Hawking had a corner office with a door he could open and close using a control on his wheelchair. Over the years, I've been fortunate enough to step through this door several times. Conversation was never easy – it took Hawking time to use his voice synthesizer – and, as I awaited an answer, I always found myself gazing around at all the paraphernalia that surrounded us. It was a remarkable array of mementos, photographs, papers, books, potted plants and what looked like general bric-a-brac, although, as I would discover, each object had a remarkable story to tell.

And why, I used to find myself thinking, was his couch so big? What was that invitation to a time traveller's party all about? And what did all those scribbles and characters on his blackboard mean? Other objects were more straightforward. When I visited in 2001, a sign next to a Lisa Simpson keyring dangling from his computer declared, "Yes, I AM the Centre of the Universe."

SO MANY QUESTIONS
Each visit, I was itching to bombard him with questions about this extraordinary trove, but maintained a respectful silence so he could focus on the matter at hand, whether we were discussing his science, an exhibit at the Science Museum or serializing one of his books.

ABOVE The nameplate of Professor Hawking's office at the Centre for Mathematical Sciences in Cambridge. Aside from scientists, film and TV crews – along with celebrities – entered through this door.

Following Stephen Hawking's death in March 2018, the contents of his office were eventually acquired for the nation, and in May 2021 allocated to the Science Museum Group. The thousand or so objects that are now in our collections, from scientific bets signed with Hawking's thumbprint to photographs of his scientific heroes, are tantamount to a personal museum curated by Stephen Hawking himself. Among them there is a rare copy of his PhD thesis, his last wheelchair, the synthesizer that gave him voice, and one of his most treasured possessions: that blackboard filled with strange doodles and arcane jokes.

Through these objects, curated by my colleague Juan-Andres Leon, you can explore his most influential ideas, from the "no-boundary" universe to how black holes leak "Hawking radiation" and his quest to "know the mind of God".

Join me, as I enter Stephen Hawking's office to explore the mind and life of one of the world's most inspirational scientists.

ABOVE The interior of Professor Hawking's office before the contents were removed to the Science Museum Group's collections.

Digitizing Hawking's Office

This ghostly "point cloud" shows how the Science Museum preserved Stephen Hawking's office in digital form so that it can be explored by online audiences.

Dave Patten and John Stack of the Museum worked with the company ScanLAB Projects to capture a digital record of his office and its contents, using high-resolution photography by means of two cameras and infrared lasers.

To create this unique record six scans were taken, each taking around a quarter of an hour. "The 3D scanning allowed us to capture the office and its contents before the objects were all packed up and moved to the Science Museum," said Patten. "The scans will allow us to recreate the office digitally in all sorts of exciting ways, as well as help us recreate the office physically at the museum at some point in the future."

Using a process known as light detection and ranging, or LiDAR, thousands of measurements were made with lasers to map objects and the scene, while the colours in the digital scan were captured using digital cameras in a process called photogrammetry. To ensure the measurements and captured colours were accurate, it was important to have the right level of lighting within the room, which was achieved by dividing it into two zones, to prevent any shadowing, and using external "lighting balloons" outside the office's windows.

Our curator of physics, Juan-Andres Leon, and an assistant digital curator, Rachael Mascarenhas, have been working with ScanLAB Projects to produce videos, walk-throughs and panoramic views to help virtual audiences explore the office for themselves. They can move from the kitchen, for example, where shelves were used to keep some of Hawking's books to the windowsills, adorned with his awards, or inspect the graffiti-strewn blackboard that kept a remarkable record of an international conference he organized in Cambridge.

In an earlier project with ScanLAB projects the museum made a virtual version of its Shipping Gallery, open from 1963 until 2012, so future generations could explore this two-tiered exhibition of hundreds of exhibits, along with many dioramas, chronicling maritime technology across time and space.

ABOVE Using a process known as light detection and ranging, or LiDAR, thousands of measurements were made with lasers to map objects and the scene, while the colours in the digital scan were captured using digital cameras in a process called photogrammetry.

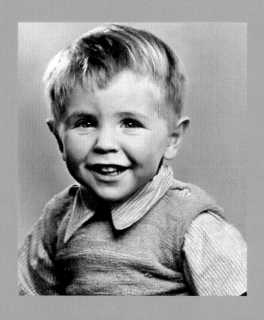

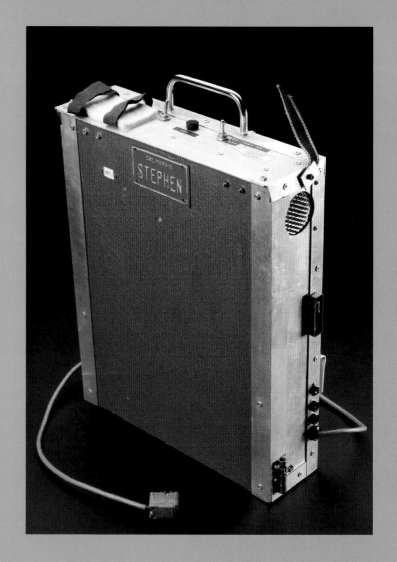

1
Early Life
and Work

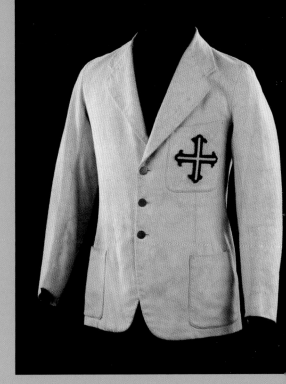

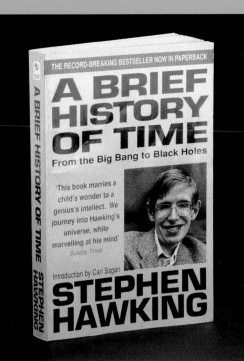

ABOVE Book with history of St Albans School, where Hawking was educated: *Born Not For Ourselves: The Story of St Albans School* by Nigel Watson, 2014.

His sister Mary recalled how as a schoolboy Stephen had a passion for model trains, was interested in finding out how things worked, and was always one for puzzles.

Young Hawking

Stephen Hawking's love of science and mathematics can be traced back to his days in St Albans, a cathedral city north of London. His formative years at school there are contained in this book, and its presence in the office underlines their significance to him.

"THEY COULD NOT SHUT HIM UP"
Born during the Second World War, in 1942, to the research biologist Dr Frank Hawking and his wife Isobel, Stephen Hawking had moved from Highgate, London, to St Albans in 1950. "In Highgate our family had seemed fairly normal," he subsequently wrote in his memoir, "but in St Albans I think we were definitely regarded as eccentric."

For a short time, Hawking attended St Albans High School for Girls, which was then open to boys until the age of ten. By that time, he had two sisters, Mary and Philippa (his parents would also adopt Edward, when Stephen was 14). In the 2021 documentary, *Hawking: Can You Hear Me?*, his sister Mary recalled how as a schoolboy Stephen had a passion for model trains, was interested in finding out how things worked, and was always one for puzzles. "Stephen was always looking for the rules that would let him win."

Another Hawking trait was clear by then. He loved being the centre of attention. Mary described how, aged eight, he was invited up on stage to sing in a pantomime in Golders Green. He belted out a long song and "they could not shut him up."

Hawking then attended St Albans School, where his classmates called him "Einstein" and with friends he would discuss many things, not least the origin of the universe.

In his last two years of school, Hawking wanted to specialize in mathematics and physics, even though he thought physics was boring "because it was so easy and obvious", and his father thought both pursuits a waste of time compared with medicine, which Stephen's sister Mary would go on to study, then practise.

Hawking was far from being a perfect pupil. He was late to learn to read, his handwriting was bad, and he could be lazy. Fortunately, his teacher Dikran Tahta proved an inspiration. "Many teachers were boring," Hawking recalled. "Not Mr Tahta. His classes were lively and exciting. Everything could be debated. Together we built my first computer: it was made with electro-mechanical switches... Behind every exceptional person", he added, "there is an exceptional teacher. Thanks to Mr Tahta, I became a professor of mathematics at Cambridge, in a position once held by Isaac Newton."

SEE ALSO:

On the Shoulders of Lucasian Giants, p.62

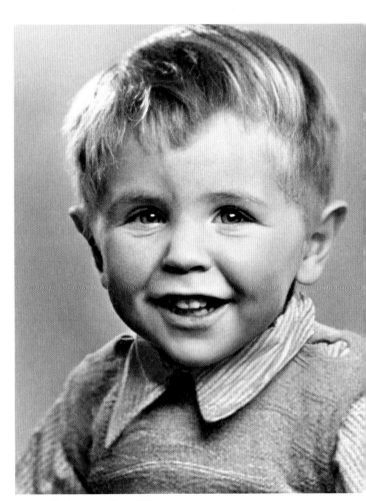

ABOVE Hawking, aged four, 1946.

Physics and High Jinks

Stephen Hawking started out his academic career as an able-bodied student and became a rowing coxswain in his third year, thanks to his strong voice, light weight and daredevil attitude.

Although he could not have been aware of it at the time, his years as an undergraduate at University College, Oxford, between 1959 and 1962 marked the end of the healthiest chapter of his life. It was only during his final year (the year the photograph shown below was taken) that he noticed he was getting clumsy and struggling with mundane tasks, such as tying his shoelaces.

After falling down some stairs, the doctor told him to "lay off the beer" and he was only eventually diagnosed with motor neurone disease after his twenty-first birthday in 1963, by which time he was a graduate student at Cambridge.

Motor neurone disease is when the nerve cells (neurones) that control muscles degenerate and die. (MND is the generic term in the United Kingdom; in the United States Amyotrophic Lateral Sclerosis, ALS, is used more widely.)

ABOVE Photograph of Hawking at his Oxford Graduation in 1962, which has degraded with exposure to sunlight.

ABOVE A portrait of Hawking (front centre) and fellow students during his undergraduate years at Oxford.

FAIRLY DISASTROUS

Stephen had agreed to apply to his father's alma mater, University College, Oxford, but reluctantly so because he wanted to study mathematics and the college had no fellow in the subject. However, his father was convinced that there would be few career prospects for mathematicians and argued that Stephen should read medicine instead.

They compromised on a natural science, physics, and Stephen went up to Oxford in 1959, at the age of 17, which even his school headmaster thought was too young. He was lonely during his first two years, but his social life blossomed during his third, not least because he joined the University College Boat Club. He was an enthusiastic participant in its high jinks, though he admitted in his memoir that "my coxing career was fairly disastrous", a reference to a disqualification and various collisions.

Among the items that he kept as mementos of his halcyon student days was his coxswain jacket. There is a note attached to the blazer to say he wore the cream-

coloured jacket when he was thrown into the river, an Oxbridge victory tradition that only a year later would have been inconceivable for him to perform.

Office curator, Juan-Andres Leon, commented: "The note that tells the story was pinned on decades later by his assistant at the time, Jeanna Lee York, who asked him if the jacket should be washed. He said: 'no!'"

NOT ANOTHER GREY MAN

By Hawking's own estimate, he had averaged an hour of work a day as an undergraduate at Oxford, in keeping with the anti-work attitudes of his peers, who regarded anyone diligent enough to get a better class of degree as a "grey man". Hawking did not prepare properly for his finals, slept poorly the night before and studiously avoided any question that demanded factual knowledge.

Even so, he performed well enough in his final examinations to be called for a "viva" (an oral examination) to determine which side of the divide between first and second class degree he lay.

When asked about his future, he told the examiners that if they awarded him a first-class degree, he would leave Oxford and go to Cambridge but if he got a second, he would stay in Oxford. They duly gave him a first as, of course, he hoped they would.

After receiving his first-class BA degree in physics, he travelled during the long summer vacation to Iran with a friend, encountering an earthquake in Bou'in-Zahra that claimed the lives of 12,000 people. Though he was near the epicentre, he did not notice the earthquake as he was ill with dysentery and bouncing around in a bus at the time, even breaking a rib as he was thrown against a seat. Safely home, he began his graduate work at Trinity Hall, Cambridge, in October 1962.

Life in Cambridge began with disappointment: he had wanted to work with Fred Hoyle, the most famous British astronomer of the day. Instead, he ended up with Dennis Sciama, "of whom I had not heard." However, his memoir added that, in retrospect, "it was probably for the best."

OPPOSITE Oxford coxing blazer, with University College badge, worn by Stephen Hawking as a student and when he was thrown into the Cherwell river, a tradition after winning.

ABOVE Hawking (far right) as a cox for the University College rowing team, Oxford.

By Hawking's own estimate, he averaged an hour of work a day as an undergraduate at Oxford, in keeping with the anti-work attitudes of his peers, who regarded anyone diligent ... as a "grey man".

Hawking's PhD Thesis

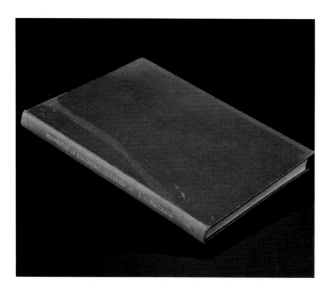

This modest book with its green cloth binding may look somewhat underwhelming. However, this PhD thesis is Stephen Hawking's intellectual equivalent of the Big Bang, and all the more remarkable because he thought he might never finish writing it. What a testament to his cosmic talent and steely determination.

The thesis, "Properties of Expanding Universes", was typed by his new wife and submitted while Hawking was a 23-year-old graduate student at Trinity Hall, Cambridge, and provides an intriguing glimpse of what was going through his mind during a time that was as difficult for him as it was exciting. "When Hawking completed his PhD, early in 1966, several leading theoretical physicists already regarded him as an outstandingly promising researcher," comments Graham Farmelo, Hawking's biographer.

His thesis topic was cosmology, the study of the origins and development of the entire universe, "widely regarded by hard-core physicists at that time as a discreditable and flaky subject", says Farmelo. "Because there were so little data to test new theories. Hawking was well aware of cosmology's reputation," he goes on, "but he set his heart on studying it, confident that it would soon evolve into a mature branch of science."

BIG BANG THEORY

Hawking's thesis would provide crucial support for the Big Bang theory, a pejorative name coined by the influential English astronomer Fred Hoyle, who argued that its proponents were somehow influenced by the Book of Genesis, seeking a moment of creation, and perhaps even a divine creator. (Hoyle himself favoured an older and more established idea which he had put forward in 1948, together with Hermann Bondi and Thomas Gold, known as the Steady State model. According to this theory, rather than stemming from a moment of creation, the universe has been around forever, and new stars are seeded in the gaps created as the universe expands.) It is a tribute to Hawking and his colleagues who popularized the notion of the Big Bang that nowadays this ultimate moment of creation is so pervasive and seems so logical to us. How could anyone have ever thought otherwise?

"When he began his PhD in autumn 1962," says Farmelo,

> he was determined to tackle only big cosmological questions whose answers were likely to change the way we think about the cosmos. Despite the diagnosis of his illness – likely to be fatal within a few years – and his research supervisor Dennis Sciama's gentle warnings about aiming too high, Hawking refused to compromise.

Hawking's thesis set him on the path to becoming the most famous scientist in the world.

Together, Penrose and Hawking showed that Einstein's general theory of relativity predicts the existence of singularities in the birth of the universe, or black holes.

INTRODUCTION

The idea that the universe is expanding is of recent
origin. All the early cosmologies were essentially
stationary and even Einstein whose theory of relativity is
the basis for almost all modern developments in cosmology,
found it natural to suggest a static model of the universe.
However there is a very grave difficulty associated with a
static model such as Einstein's which is supposed to have
existed for an infinite time. For, if the stars had been radi-
ating energy at their present rates for an infinite time,
they would have needed an infinite supply of energy. Further,
the flux of radiation now would be infinite. Alternatively,
if they had only a limited supply of energy, the whole universe
would by now have reached thermal equilibrium which is certainly
not the case. This difficulty was noticed by Olbers who however
was not able to suggest any solution. The discovery of the
recession of the nebulae by Hubble led to the abandonment of
static models in favour of ones which were expanding.

Clearly there are several possibilities: the universe
may have expanded from a highly dense state a finite time ago
(the so-called 'big-bang' model); another is that the present
expansion may have been preceded by a contraction which, in

OPPOSITE Hawking's doctoral thesis, entitled "Properties of Expanding Universes". **ABOVE** A page from the thesis.

ABOVE Trinity Hall, Cambridge. Hawking studied a PhD in Cosmology at Trinity Hall between 1962 and 1965.

"MY OWN ORIGINAL WORK"

However, only a handful of copies were made. For example, there is one, signed and dated 15 October 1965, which contains a typed dedication to Dennis Sciama, an influential physicist in his own right. All copies contain handwritten equations, because typewriters couldn't manage mathematical formulae, so these had to be added later, probably by his wife, Jane. Additionally, inside some copies (such as the one that can be inspected online), one gets a rare glimpse of Hawking's handwriting: a note scribbled in pencil reads, "This dissertation is my original work – SW Hawking".

The year after he began the thesis Hawking was diagnosed with the motor neurone disease that doctors predicted would kill him within a few years. There was a chance he might never finish it. Nevertheless, he was determined to continue with his doctorate, partly so that he could get a job afterwards to support Jane.

"In his first two years as a postgraduate student, Hawking investigated several worthwhile topics, and demonstrated that he was a promising talent," says Farmelo.

His research only took off in his final year, however, when he heard about an exciting new discovery by the well-established mathematical physicist Roger Penrose. In a brilliant insight, Penrose had proved mathematically that Einstein's theory of gravity predicts that sufficiently heavy stars must collapse into a cosmic object, later dubbed "a black hole", that contains a singularity.

When Penrose set out his discovery in a seminar at King's College London, in January 1965, Hawking heard about it from Brandon Carter, a Cambridge colleague who had attended. Then Sciama made a point of bringing them together and, as Penrose recalled, it did not take Hawking long to find a way to use his theorem unexpectedly. "They got on well, became friends and later collaborated on classic papers about the mathematical singularities in Einstein's theory of gravity," says Farmelo.

SINGULARITY THEORIST

Hawking applied Penrose's ingenious technique to the entire universe. "This application of Penrose's ideas to the universe attracted the attention of leading theorists, many of whom recognized Hawking's potential," explains Graham Farmelo. "He was soon widely recognized as one of the leading young gravity theorists to stand on the shoulders of Penrose, and later established himself as a singularly gifted theorist in his own right."

Together, Penrose and Hawking showed that Einstein's general theory of relativity predicts the existence of singularities in the birth of the universe, or black holes. A singularity is the point where gravity becomes infinite and space-time, the union of time and space used in relativity, simply ends. As the presence of infinities in these singularities suggests, in such extreme situations the laws of physics themselves break down, and Penrose and Hawking hinted that a deeper physics was required.

THE PHYSICS OF BLACK HOLES

Hawking would become the preeminent expert on the physics of black holes. Cygnus X-1, a galactic X-ray source discovered in 1964 in the constellation Cygnus – visible in the northern hemisphere's summertime sky – was the first cosmic object believed to be a black hole, and became the subject of a famous bet between Stephen Hawking and his collaborator and future Nobel Prize winner Kip Thorne. As Hawking put it in his book, *A Brief History of Time*, he bet against the existence of black holes in the region as an "insurance policy". "I have done a lot of work on black holes," he wrote, "and it would all be wasted if it turned out that black holes do not exist. But in that case, I would have the consolation of winning my bet, which would win me four years of the magazine *Private Eye*." In his own book, *Black Holes and Time Warps*, Thorne described how in 1990 Hawking eventually admitted he had lost the bet by breaking into Thorne's office late one night with his entourage while Thorne was in Russia, finding the framed bet, and leaving a note that conceded, "signed" with his thumbprint.

Decades later, in the spring of 2019, astronomers

unveiled the first image of a black hole: M87 at the centre of the Virgo galaxy, a behemoth with a mass 6.5 billion times that of the Sun, located 54 million light years away from Earth. The feat was announced by a global research team of 300 researchers from 80 institutes, known as the Event Horizon Telescope Collaboration, which used observations from a worldwide network of radio telescopes. Although we cannot see the black hole itself, the image shows a tell-tale signature: a dark central region (called a "shadow") surrounded by a bright, ring-like structure. The new view captured light bent by the vast gravitational tug of the black hole at our galaxy's heart, which is four million times more massive than our own Sun.

In the wake of this proof of the existence of black holes, Roger Penrose, now at the University of Oxford, was awarded the 2020 Nobel Prize for Physics, "for the discovery that black hole formation is a robust prediction of the general theory of relativity." He shared it with Reinhard Genzel of the Max Planck Institute for Extraterrestrial Physics in Garching, Germany, and the University of California, Berkeley, and Andrea Ghez of the University of California in Los Angeles, for their discovery of a supermassive compact object at the centre of our galaxy. The only known explanation for this object was that it was indeed a black hole.

If only Hawking had been alive, he might have shared the prize with Penrose. "Hawking longed to win the Nobel Prize for Physics, and rarely missed an opportunity to say so," says Graham Farmelo:

Despite several nominations, the prize eluded him. The theory is that the Nobel authorities, slow to recognize progress in gravity theory, did not award their prize for these exotic objects until astronomers confirmed their existence beyond all reasonable doubt. When Penrose deservedly won the prize for his seminal work on understanding black holes, Hawking had been dead for two-and-a-half years and was therefore ineligible for the award. His much trumpeted ambition was, sadly, unfulfilled.

SEE ALSO:
Hawking Radiation, p.46
Stephen Hawking, You Are Wrong!, p.176

ABOVE X-ray telescope image of the binary star system known as Cygnus X-1, a black hole "candidate".

ABOVE Stephen and Jane Hawking's wedding photo from the cover of *Blanco Negro* magazine, 22 October 1989, which featured an interview and article on Stephen Hawking: "un genio en la intimidad" (a genius in privacy).

A Brief Guide to Hawking's Thesis

The abstract of Stephen Hawking's doctoral thesis begins, modestly, "Some implications and consequences of the expansion of the universe are examined." But those implications and consequences go to the very heart of our understanding of space and time. The thesis was approved in February 1966, after a viva voce, or "viva", an oral test during which he had to defend his thesis before his peers.

What was it all about? In his research, Stephen Hawking was building on the work of others, working through the implications of Albert Einstein's theory of gravity, his general theory of relativity, presented in November 1915. While Isaac Newton saw gravity as a tug between two objects, Einstein envisaged it in a radically different way, in terms of warping the fabric of space and time (space-time) around objects. The theory provided an entirely new foundation for understanding gravity, which shapes the universe at the largest scale.

The first theoretical description of what we now call a black hole came just a few weeks after the publication of Albert Einstein's general theory of relativity, when the German astrophysicist Karl Schwarzschild described how heavy masses bend space and time. The idea of black holes, in the guise of what were called "dark stars", was first considered at the end of the eighteenth century, in the works of the British philosopher and mathematician John Michell and the renowned French thinker Pierre-Simon de Laplace. In an impressive display of prescience, both had reasoned that heavenly bodies could become so dense that they would be invisible – not even the speed of light would be fast enough to escape their gravity.

Up until the 1960s, Schwarzschild's black hole solutions were regarded as theoretical speculation, describing ideal situations in which stars and their black holes were perfectly round and symmetrical. Whether black holes could actually form under realistic conditions was finally resolved in 1964 by Roger Penrose, who was then a professor of mathematics at Birkbeck College, London.

His idea, which he called trapped surfaces, would be a crucial mathematical tool needed to describe a black hole. These surfaces are formed by the explosion of a star at the end of its life, thus causing its collapse and the subsequent formation of a black hole. A trapped surface forces all rays to point towards a centre, regardless of whether the surface curves outwards or inwards. Using the concept of trapped surfaces, Penrose was able to mathematically prove that a black hole always conceals a singularity, a place where the usual laws of physics cease to apply. He published his work in January 1965, ten years after Einstein's death, and it marked the most important contribution to the general theory of relativity since Einstein himself.

THE UNIVERSE EXPLODING INTO BEING

As the title of his thesis, which references Penrose's work, suggests, Stephen Hawking applied the mathematics of general relativity to models of the birth of our universe. As he explained in *A Brief History of Time*, he ran the theory by which black holes could be described in reverse, showing that the universe must have exploded into being from a singularity, a single point in space and time. His work would have profound consequences and, as is clear from a couple of sections crossed out in the submitted thesis, Hawking was wrestling with these ideas and their implications right up to the moment he submitted it.

The prevailing theory about the creation of the universe, known as the Steady State theory, saw the universe not as issuing from a Big Bang but as a static entity that had existed forever. It had been encapsulated in a model called Hoyle-Narlikar theory, the Hoyle in question being Britain's best-known astronomer, Sir Fred Hoyle. The Steady State theory had already been brought into question by the observations of the American astronomer Edwin Hubble, which suggested that the universe was expanding and, by inference, that it must have had a beginning – a moment of creation, or Big Bang. But it remains a remarkable testament to Stephen Hawking's intellect (and cockiness) that he was going to take on Hoyle by being one of the first to say that the theory was not correct.

In his thesis, Hawking laments that, although the general theory of relativity is powerful, it allows for many different solutions to its equations. The Steady State model fails to match observations, not least what is called the Friedmann-Lemaître-Robertson-

Walker metric, a solution to Einstein's equations which assumes that matter is evenly distributed in an expanding (or contracting) universe – a premise backed up by astronomical observations.

PERTURBATIONS AND GRAVITATIONAL WAVES

Hawking's second chapter covers perturbations – small variations in the local curvature of space-time – and how they evolve as the universe expands, concluding that a tiny perturbation "will not contract to form a galaxy". We now know that this idea is wrong, and that dark matter, a mysterious source of gravity in the universe, does seed galaxy formation.

Hawking also discusses gravitational waves – ripples in the fabric of space-time that move through the universe. These were by then becoming a fashionable area of study, though they were only actually detected for the first time in September 2015 using the Laser Interferometer Gravitational-Wave Observatory (LIGO), to which Hawking's old friend, Kip Thorne, made crucial contributions.

Before reaching its striking conclusion, Hawking's thesis considers the shape of space-time and whether the universe is "closed", "open" or "flat". This has implications for our understanding of the expansion of the universe, as in a closed shape the density of the universe is either so high that it collapses into a big crunch, or so low that it expands forever, or delicately adjusted to the extent that, eventually, it would neither expand nor contract.

Hawking goes on to prove that space-time can begin and end at a singularity, or Big Bang and Big Crunch: He breaks down the assumption that "space-time is singularity-free" by showing that such a universe would be simultaneously both open and closed. "This", he writes, "is a contradiction" – and science does not like contradictions. Hawking ends the penultimate paragraph of his thesis with this conclusion: "Thus the assumption that space-time is non-singular must be false."

THE UNIVERSE HAD A BEGINNING

In his final chapter, and by far the most important, Hawking applies Roger Penrose's theorem of a space-time singularity in the centre of black holes and to the entire universe. It enables him to show that it is possible for space-time to begin as a singularity – that space and time in our universe did have an origin. Here at last was mathematical proof to confirm that the universe must have had a beginning. Over the coming decades observations of the universe would verify Hawking's work, while studies of the cosmic microwave background – the leftover echo from the Big Bang – would finally bury the Steady State model once and for all.

When Hawking submitted his thesis in 1966, he could never have imagined the popular appeal it would one day have. Half a century later, he had enthusiastically given his consent to have it made open-access – that is, made available free of charge to the public. "I hope to inspire people around the world to look up at the stars and not down at their feet," he said in 2017;

> to wonder about our place in the universe and to try and make sense of the cosmos ... It's wonderful to hear how many people have already shown an interest in downloading my thesis – hopefully they won't be disappointed now that they finally have access to it!

When the high demand for printed copies – even priced at $85 – prompted the University of Cambridge to make Hawking's dissertation available online, interest was so intense that the download site crashed. Stephen Hawking's thesis had broken the Internet.

ABOVE Hawking with Andrew Wiles, Martin Bridson and Roger Penrose at the Mathematical Institute, University of Oxford.

physicsworld

physicsworld.com

Volume 31 No 1 January 2018

COSMIC IMPACT

Colliding neutron stars send
ripples through physics

Working in the dark What life is like for blind physicists
Reactor control Using ultrasound to monitor Fukushima
Scientific limits Why academia shouldn't ignore the arts

Hawking Makes Waves

Gravitational waves are ripples in space and time that provide scientists with a new way of studying the universe.

Strong and detectable waves are produced by cataclysmic cosmic events such as colliding black holes, supernovae (massive stars exploding at the end of their lifetimes) and colliding neutron stars. Predicted in 1916 by Albert Einstein's theory of general relativity, a century before they were detected, they had fascinated Stephen Hawking since the 1960s. Remarkably, his interest in these cosmic ripples almost diverted him from theoretical work towards experiments to find practical ways to detect them. Fortunately for cosmology, his bid to become an experimenter proved fruitless.

When objects accelerate, they produce gravitational waves travelling at light speed, but to make waves big enough to detect takes massive objects such as pairs of black holes. As they orbit each other, they emit gravitational waves that gradually drain away some orbital energy to the point where the black holes themselves eventually collide, releasing a further, this time giant, burst of gravitational waves.

The hope among scientists in the 1960s was that they could build observatories to detect such ripples, which are tens to hundreds of miles long from crest to crest and roll past Earth in fractions of a second. Their detection would be momentous, confirming Einstein's general theory of relativity and throwing open an unprecedented new window onto the most violent cosmic events.

OPPOSITE *Physics World* magazine issue from early 2018 with articles on the first-ever observation of gravitational waves from two neutron stars as they spiralled into each other, then merged to form a black hole.

The interest of Hawking's peers in these waves was piqued when in 1969 the American physicist Joe Weber claimed to have detected gravitational radiation from the centre of our galaxy, using what was known as a bar-type detector. "He had a large aluminium bar," recalled Gary Gibbons, Hawking's student around that time, "which was being set in oscillation by the gravitational waves, which alternatively stretched and squeezed it, or so he said."

In 1970 Hawking visited Weber, who lived near Princeton, and subsequently recalled Weber's findings in his memoir. "When a gravitational wave came along," wrote Hawking,

> it would stretch things in one direction (perpendicular to the direction of travel of the wave) and compress things in the other direction (perpendicular to the wave). This would make the bars oscillate at their resonant frequency – 1,660 cycles per second – and these oscillations would be detected by crystals strapped to the bars.

Weber's work was greeted with some scepticism, but it did stimulate a lot of interest. Soon afterwards Hawking had a visit in Cambridge from Peter Aplin from the University of Bristol, who was thinking about how to boost the sensitivity of bar-type detectors to make their data more convincing.

This is when Gary Gibbons would first cross paths with Stephen Hawking. Though they had seen each other before in the tea room in Silver Street, where the Department of Applied Mathematics and Theoretical Physics (DAMTP) was then based, their discussion in the wake of Aplin's visit marked the first time they had met.

The hope among scientists in the 1960s was that they could build observatories to detect such ripples, which are tens to hundreds of miles long from crest to crest and roll past Earth in fractions of a second.

"Stephen and I got talking about gravitational wave detection," remembered Gibbons. Most of the people in their department were mathematicians and did not have much intuition about building instruments, he recalled, but "Stephen got the idea of actually detecting gravitational waves in our laboratory. This was a rather silly idea, but we pursued it for a while."

To grasp how easy it would be to detect the waves, in 1971 Hawking calculated an upper limit on the energy given off by colliding black holes (we now know that this is about 20 times higher than the energy actually given off). In the same year, Stephen co-authored a paper with Gary Gibbons on the possibility of such a cosmic event being detected. Encouraged by both their old supervisor Dennis Sciama and George Batchelor, the head of the DAMTP at the time, Stephen put in an application for funds to develop an improved detector, which would be constructed in the basement of the department. He then withdrew it, which was just as well. "That was a narrow escape!" Hawking recalled. "My increasing disability would have made me hopeless as an experimenter ... I'm very glad I remained a theorist."

THE RISE OF LIGO

Although Weber's design continued to be developed, in the 1970s scientists came to realize that other approaches to detecting gravitational waves might be more productive, notably bouncing a laser beam between free-hanging mirrors. The best-known observatory carrying out such research today, the Laser Interferometer Gravitational Wave Observatory (LIGO), consists of two vast L-shaped observatories in America, one in Washington state and the other in Louisiana.

The two of them were built 3,000 kilometres apart to ensure they would actually detect the elusive signal: if something was detected by only one of them, it might not be from a gravitational wave. Detectors situated far apart, however, will not sense the same local vibrations, but will still detect the passage of a gravitational wave at almost the same time. Each observatory sends laser beams down two 4 kilometre-long arms at right angles to each other, which bounce off mirrors at each end and are then re-combined near the source (a prototype LIGO beam splitter is in the Science Museum Group collection). If a gravitational wave were to pass through, it would "wiggle" the fabric of space-time and move the mirrors, causing a tiny difference between the arrival times of the peaks and troughs of the waves of the laser beams, and produce what is known as an interference pattern.

LIGO became operational in 2001, and was upgraded in May 2015. Eventually, in September 2015 LIGO picked up the first ever direct signature of gravitational waves, sent out by the merger of two black holes. As a result, the 2017 Nobel Prize for Physics was shared by the LIGO pioneers Kip S. Thorne, Hawking's old friend whom he had known since 1965, Rainer Weiss and Barry C. Barish.

"Stephen got the idea of actually detecting gravitational waves in our laboratory. This was a rather silly idea, but we pursued it for a while."

ABOVE The Laser Interferometer Gravitational-Wave Observatory (LIGO) gravitational wave detector, which operates on two sites, one near Hanford in eastern Washington, and another near Livingston, Louisiana, shown here.

INCREDIBLE PRECISION

The waves caused by the joining of two black holes "shake" space as they pass through the Earth, but with an amplitude so minuscule that, as the Astronomer Royal, Lord Rees, commented, to pick up this signal was "like detecting the thickness of a human hair at the distance of [the distant star system] Alpha Centauri." To do this required ten million times more precision than Joe Weber's bar-type detector could offer. At its most sensitive, LIGO could detect a change in distance 1/10,000th the width of a proton.

This incredible precision also offered a remarkable way to experimentally validate Stephen Hawking's 1971 area theorem, stimulated by an idea of Roger Penrose's, that the total area of a black hole can never decline, a profound insight which catalysed a series of fundamental breakthroughs in the thinking about black hole thermodynamics. On hearing about the 2015 LIGO results, Hawking approached Kip Thorne to see if the technique could confirm his notion.

At the time, researchers did not have the ability to sift the necessary information within the signal, before and after the merger of the black holes, to determine whether Hawking's calculations were correct. But in 2019 Maximiliano Isi, a NASA Einstein Postdoctoral Fellow at MIT's Kavli Institute for Astrophysics and Space Research, and his colleagues were able to take a closer look at GW150914, the first gravitational wave signal created by merging black holes. If Hawking's area theorem held true, then the horizon area of the new black hole should not be smaller than the total combined horizon area of its parent black holes. With the aid of computer simulations, Isi and his colleagues developed a way to analyse the gravitational wave signals before and after the two parent black holes collided, and determine whether the final horizon area had increased.

They re-analysed the signal from GW150914 before and after the cosmic collision, and in 2021, three years after Hawking's death, reported that the results were indeed consistent with what he had predicted. They managed to identify the mass and spin of both black holes before they merged and, from these estimates, calculated their total horizon areas to be roughly 235,000 square kilometres. Then, from the reverberations of the newly formed black hole, they calculated its mass and spin, and ultimately its horizon area: 367,000 square kilometres. There were uncertainties around these numbers, due to noise in LIGO's data, but even so, the team could declare with 97 per cent confidence that the area had verifiably increased.

Although Hawking did not live to see this evidence that his black hole area theorem was correct, his work had shown other ways in which gravitational waves can give deep insights into the universe. With the development of the theory of inflation, which says that within its first trillionth of a second the universe ballooned from sub microscopic to astronomical size, it became clear that gravitational waves should have been emitted at the same time, given that inflation was such a violent cosmic event.

In 2000, Stephen, together with his former student Thomas Hertog and South African physicist Neil Turok, used methods closer in spirit to those developed with Gibbons in the 1970s to calculate the spectrum of these primordial gravitational waves.

HAWKING'S FINAL THEORY

In some of his last work on the origin of the universe, with Hertog, Stephen Hawking used string theory to predict that the universe is finite, and far simpler than many theories predict. In some regions of the universe, he argued, quantum effects can keep inflation going for ever so that, overall, inflation is eternal. If this were so, then the observable part of our universe would be just a hospitable "pocket universe": a rare region where inflation has ended, and stars and galaxies form, in a mosaic of pocket universes, or multiverse, somewhat like bubbles in boiling water.

Hawking, however, had "never been a fan of the multiverse". "If the scale of different universes in the multiverse is large or infinite," he wrote, then "the theory can't be tested." In *On the Origin of Time: Stephen*

Although Hawking did not live to see this evidence that his black hole area theorem was correct, his work had shown other ways in which gravitational waves can give deep insights into the universe.

Hawking's final theory "is more than just a scientific cosmology ... It is a cosmology in the humanistic sense, in which the universe is seen as our home."

Hawking's Final Theory, Thomas Hertog recalled that, even at their very first meeting in 1998, Hawking had said of the multiverse: "This is outrageous." In their final paper together, Hawking and Hertog declared that this account of eternal inflation as a theory of the Big Bang is wrong. "We are not down to a single, unique universe," said Hawking, "but our findings imply a significant reduction of the multiverse, to a much smaller range of possible universes."

Their study made use of the new concept of holography in string theory, which postulates that the universe is akin to a hologram: physical reality in certain 3D spaces can be mathematically reduced to 2D projections on a surface. "In our paper we put forward a mathematical model for the state of the universe at the beginning," said Hertog. "We then use this to predict what kind of universes can come into existence." This also rendered the theory more testable. Hertog believes that primordial gravitational waves constitute the most promising way to test their model: "Generally speaking, our theory predicts there is a contribution from gravitational waves generated during inflation to the pattern of cosmic microwave background variations." At the time of his death, reveals Hertog, Stephen Hawking was working on a draft of a paper specifically devoted to testing inflation with gravitational waves. "The observation of signatures of those gravitational waves from the Big Bang would be a smoking gun we are on the right track."

The expansion of our universe since its beginning means that these gravitational waves would have very long wavelengths, currently beyond the range of LIGO. But they might be glimpsed in future experiments to measure the echo of the Big Bang, or the polarization of the cosmic microwave background, or picked up by a successor to the planned European space-based gravitational wave observatory, LISA, the Laser Interferometer Space Antenna.

A few weeks before he died, Hawking met once more with Hertog, who recalled his last words to him: "Time for a new book ... include holography." Hawking's final theory "is more than just a scientific cosmology", according to Hertog. "It is a cosmology in the humanistic sense, in which the universe is seen as our home – albeit a big one – and its physics rooted in our relationship with it."

SEE ALSO:

Hawking Radiation, p.46
Beyond the No Boundary, p.156

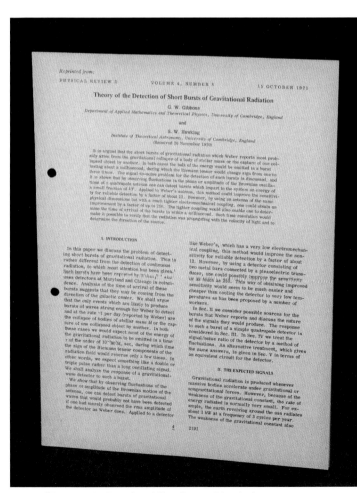

ABOVE "Theory of the Detection of Short Bursts of Gravitation Radiation", an article from the scientific journal *Physical Review D*, 1971.

Hawking Radiation

In 1974, Stephen Hawking shook the world of physics with a paper that could be summed up by an equation that, he later insisted, should be inscribed on his tombstone.

According to Einstein's theory of general relativity, absolutely nothing, including light, can escape from inside a black hole. That, of course, explains why a black hole is black. But this simple picture would change when quantum theory, which rules the subatomic world, was also applied to these extreme astronomical objects.

The first step towards Hawking's 1974 paper came in 1970 in the wake of his work with Roger Penrose on general relativity, black holes and the birth of the universe. A few days after the birth of his daughter Lucy, Hawking had a "Eureka moment". He realized "first of all how to define a black hole in a more mathematical sense than its colloquial sense", explained Gary Gibbons, his student at the time, and, using tools developed by Roger Penrose, "he defined what we call the event horizon, the boundary of the events that can be seen from infinity." It was a crucial concept: something enfolding a black hole whose extraordinary gravitational tug traps light and matter so they can never escape beyond this horizon.

Another idea of Penrose's set Hawking thinking about what happens when black holes collide, at a time when he was pondering the ripples in space-time that these violent collisions would send out. It would set the stage for his "area theorem", and trigger a chain reaction of fundamental insights into the nature of black holes, and lead to Hawking's remarkable realization that black holes are not so black after all.

His area theorem predicts that the total area of a black hole's event horizon – and all black holes in the universe, for that matter – should never decrease. Hawking noted that the ever-expanding nature of the area of the event horizon was reminiscent of entropy, which you can think of as a measure of disorder, which always increases, according to the second law of thermodynamics. Thermodynamics is the science that governs universal concepts of energy, temperature and entropy, and Hawking's theorem hinted at a subtle link between it and black holes. "That was a very striking result," said Gary Gibbons.

There also seemed to be a link between event horizons and entropy, as Hawking pointed out in a paper in 1971, but he shrugged it off as just a coincidence. The problem was that all objects that have entropy are warm and radiate heat, but in the case of black holes this struck Hawking as impossible, because general relativity suggests that nothing, including heat radiation, can escape beyond the event horizon.

Around the same time, however, this same question was being pondered by Jacob Bekenstein, a PhD student at Princeton University, along with his supervisor, John Wheeler. Wheeler had popularized terms such as black hole (previously they were called "frozen stars") and wormhole and, according to Hawking himself, was the "hero of the black hole story".

BLACK HOLES WITH NO HAIR
In 1939 Albert Einstein had written a paper which suggested that stars could not collapse under gravity, because matter could not be compressed beyond a certain

OPPOSITE Hawking's "Black Hole Explosions?" paper published in the journal Nature, in 1974, which disproved the conventional notion that nothing could ever escape from a black hole. It boldly blended three major theories: quantum, general relativity and thermodynamics.

Hawking's area theorem predicts that the total area of a black hole's event horizon – and all black holes in the universe, for that matter – should never decrease.

(Reprinted from Nature, Vol. 248, No. 5443, pp. 30–31, March 1, 1974)

Black hole explosions?

QUANTUM gravitational effects are usually ignored in calculations of the formation and evolution of black holes. The justification for this is that the radius of curvature of space-time outside the event horizon is very large compared to the Planck length $(G\hbar/c^3)^{1/2} \approx 10^{-33}$ cm, the length scale on which quantum fluctuations of the metric are expected to be of order unity. This means that the energy density of particles created by the gravitational field is small compared to the space-time curvature. Even though quantum effects may be small locally, they may still, however, add up to produce a significant effect over the lifetime of the Universe $\approx 10^{17}$ s which is very long compared to the Planck time $\approx 10^{-43}$ s. The purpose of this letter is to show that this indeed may be the case: it seems that any black hole will create and emit particles such as neutrinos or photons at just the rate that one would expect if the black hole was a body with a temperature of $(\kappa/2\pi)(\hbar/2k) \approx 10^{-6}\ (M_\odot/M)K$ where κ is the surface gravity of the black hole [1]. As a black hole emits this thermal radiation one would expect it to lose mass. This in turn would increase the surface gravity and so increase the rate of emission. The black hole would therefore have a finite life of the order of $10^{71}\ (M_\odot/M)^{-3}$ s. For a black hole of solar mass this is much longer than the age of the Universe. There might, however, be much smaller black holes which were formed by fluctuations in the early Universe[2]. Any such black hole of mass less than 10^{15} g would have evaporated by now. Near the end of its life the rate of emission would be very high and about 10^{30} erg would be released in the last 0.1 s. This is a fairly small explosion by astronomical standards but it is equivalent to about 1 million 1 Mton hydrogen bombs.

To see how this thermal emission arises, consider (for simplicity) a massless Hermitean scalar field ϕ which obeys the covariant wave equation $\phi_{;ab}g^{ab} = 0$ in an asymptotically flat space time containing a star which collapses to produce a black hole. The Heisenberg operator ϕ can be expressed as

$$\phi = \sum_i \{f_i a_i + \bar{f}_i a_i^+\}$$

where the f_i are a complete orthonormal family of complex valued solutions of the wave equation $f_{i;ab}g^{ab} = 0$ which are asymptotically ingoing and positive frequency—they contain only positive frequencies on past null infinity I^- [3,4,5]. The position-independent operators a_i and a_i^+ are interpreted as annihilation and creation operators respectively for incoming scalar particles. Thus the initial vacuum state, the state containing no incoming scalar particles, is defined by $a_i|0_-\rangle = 0$ for all i. The operator ϕ can also be expressed in terms of solutions which represent outgoing waves and waves crossing the event horizon:

$$\phi = \sum_i \{p_i b_i + \bar{p}_i b_i^+ + q_i c_i + \bar{q}_i c_i^+\}$$

where the p_i are solutions of the wave equation which are zero on the event horizon and are asymptotically outgoing, positive frequency waves (positive frequency on future null infinity I^+) and the q_i are solutions which contain no outgoing component (they are zero on I^+). For the present purposes it is not necessary that the q_i are positive frequency on the horizon even if that could be defined. Because fields of zero rest mass are completely determined by their values on I^-, the p_i and the q_i can be expressed as linear combinations of the f_i and the $\bar{f}_i$:

$$p_i = \sum_i \{\alpha_{ij}f_i + \beta_{ij}\bar{f}_i\} \quad \text{and so on}$$

The β_{ij} will not be zero because the time dependence of the metric during the collapse will cause a certain amount of mixing of positive and negative frequencies. Equating the two expressions for ϕ, one finds that the b_i, which are the annihilation operators for outgoing scalar particles, can be expressed as a linear combination of the ingoing annihilation and creation operators a_i and a_i^+

$$b_i = \sum_i \{\bar{\alpha}_{ij}a_i - \bar{\beta}_{ij}a_i^+\}$$

Thus when there are no incoming particles the expectation value of the number operator $b_i^+ b_i$ of the ith outgoing state is

$$\langle 0_-\ |b_i^+ b_i|\ 0_-\rangle = \sum_j |\beta_{ij}|^2$$

The number of particles created and emitted to infinity in a gravitational collapse can therefore be determined by calculating the coefficients β_{ij}. Consider a simple example in which the collapse is spherically symmetric. The angular dependence of the solution of the wave equation can then be expressed in terms of the spherical harmonics Y_{lm} and the dependence on retarded or advanced time u, v can be taken to have the form $\omega^{-1/2} exp\ (i\omega u)$ (here the continuum normalisation is used). Outgoing solutions $p_{lm\omega}$ will now be expressed as an integral over incoming fields with the same l and m:

$$p_\omega = \int \{\alpha_{\omega\omega'}f_{\omega'} + \beta_{\omega\omega'}\cdot\bar{f}_{\omega'}\}\ d\omega'$$

(The lm suffixes have been dropped.) To calculate $\alpha_{\omega\omega'}$ and $\beta_{\omega\omega'}$ consider a wave which has a positive frequency ω on I^+ propagating backwards through spacetime with nothing crossing the event horizon. Part of this wave will be scattered by the curvature of the static Schwarzschild solution outside the black hole and will end up on I^- with the same frequency ω. This will give a $\delta(\omega - \omega')$ behaviour in $\alpha_{\omega\omega'}$. Another part of the wave will propagate backwards into the star, through the origin and out again onto I^-. These waves will have a very large blue shift and will reach I^- with asymptotic form

$$C\omega^{-1/2} \exp\ \{-i\omega\kappa^{-1}\ \log\ (v_0 - v) + i\omega v\}\ for\ v < v_o$$

and zero for $v \geq v_o$, where v_o is the last advanced time at which a particle can leave I^-, pass through the origin and escape to I^+. Taking Fourier transforms, one finds that for large ω', $\alpha_{\omega\omega'}$ and $\beta_{\omega\omega'}$ have the form:

$$\alpha_{\omega\omega'} \approx C \exp\ [i(\omega - \omega)v_0](\omega'/\omega)^{1/2}$$
$$\cdot\Gamma(1 - i\omega/\kappa)[-i(\omega - \omega')]^{-1+i\omega/\kappa}$$
$$\beta_{\omega\omega'} \approx C \exp\ [i(\omega + \omega)v_0](\omega'/\omega)^{1/2}$$
$$\cdot\Gamma(1 - i\omega/\kappa)[-i(\omega + \omega')]^{-1+i\omega/\kappa}.$$

The total number of outgoing particles created in the frequency range $\omega \to \omega + d\omega$ is $d\omega\int_0^\infty|\beta_{\omega\omega'}|^2 d\omega'$. From the above expression it can be seen that this is infinite. By considering outgoing wave packets which are peaked at a frequency ω and at late retarded times one can see that this infinite number of particles corresponds to a steady rate of emission at late retarded times. One can estimate this rate in the following way. The part of the wave from I^+ which enters the star at late retarded times is almost the same as the part that would have crossed the past event horizon of the Schwarzschild solution had it existed. The probability flux in a wave packet peaked at ω is roughly proportional to $\int_{\omega'}\omega^2/\{\alpha_{\omega\omega'}|^2 - |\beta_{\omega\omega'}|^2\}\ d\omega$ where $\omega_2' \gg \omega_1' > 0$. In the expressions given above for $\alpha_{\omega\omega'}$ and $\beta_{\omega\omega'}$ there is a logarithmic singularity in the factors $[-i(\omega - \omega')]^{-1+i\omega/\kappa}$ and $[-i(\omega + \omega')]^{-1+i\omega/\kappa}$. Value of the expressions on different sheets differ by factors of $\exp(2\pi n\omega\kappa^{-1})$. To obtain the correct ratio of $\alpha_{\omega\omega'}$ to $\beta_{\omega\omega'}$ one has to continue $[-i(\omega + \omega')]^{-1+i\omega/\kappa}$

According to Hawking's daughter, Lucy, the significance of a quantum perspective became clear during a warm-up mathematical exercise.

ABOVE Member of the Soviet Academy of Sciences, physicist and nuclear weaponeer Yakov Zel'dovich.

point. But in the 1950s and 1960s, Wheeler's research had emphasized that many stars *would* eventually collapse, and he "foresaw many of the properties of the objects which collapsed stars become, that is, black holes," said Hawking. No matter what is "eaten" by a black hole, from other black holes to errant spacefarers, all that will change about the black hole is its mass, charge and state of rotation. In reference to this, Wheeler liked to say that a black hole has "no hair" – a metaphoric way of saying that it is tricky to tell black holes apart because only their mass, charge and spin can distinguish them: they have no other characteristics – unlike our Sun, for example, which is not perfectly spherical, and has a complex gravitational field, sunspots and solar prominences. (In 1972 Stephen Hawking's colleague Brandon Carter, along with supporting work by Hawking himself and David Robinson at King's College London, did indeed demonstrate that "a black hole has no hair" when they showed mathematically, based on various assumptions, that a black hole is described by just three parameters).

ENTER BEKENSTEIN
To tease out the link with thermodynamics, Jacob Bekenstein dreamed up a little thought experiment: imagine what happens if a box of hot gas crosses the event horizon. If one argues that nothing escapes, then both the box of gas and the entropy associated with it vanish from our universe, which directly contradicts the second law of thermodynamics.

Bekenstein worked out that increasing the entropy of a black hole increased its mass. On one hand, this was in line with Stephen Hawking's paper that showed how event horizons never shrink. But on the other hand, it challenged Hawking, because his work suggested that the area of the event horizon of a black hole was not an analogy but a direct measure of entropy. The entropy of the universe always increases, even when things fall into

black holes, because the entropy lost from the space outside the event horizon is compensated by the rise the surface area of the event horizon. This, said Bekenstein, was a generalized form of the second law of thermodynamics. Indeed, it can be seen as a cosmic formulation of this famous law of science.

Bekenstein's paper on this entropy appeared in 1972, but ignored the implication that a black hole had to radiate heat. No one, let alone Hawking, believed that was possible. In *A Brief History of Time*, Hawking wrote that it made him want to take a closer look, since he felt Bekenstein had misused his own discovery of the increase of the area of the event horizon.

ENTER ZEL'DOVICH
But Hawking would revise his thinking after his old friend, Kip Thorne at Caltech, organized a trip to Moscow in September 1973 to meet a great Soviet astrophysicist, Yakov Borisovich Zel'dovich. Zel'dovich was keen to meet Stephen but, as one of the fathers of the Soviet hydrogen bomb, was forbidden during the Cold War from travelling to Cambridge. Stephen had "wowed Zel'dovich", according to Thorne, and in turn, Zel'dovich and his PhD student Alexei Starobinsky had, during a meeting at the Rossiya Hotel, inspired Stephen with their discovery that a spinning black hole can lose energy by creating particles.

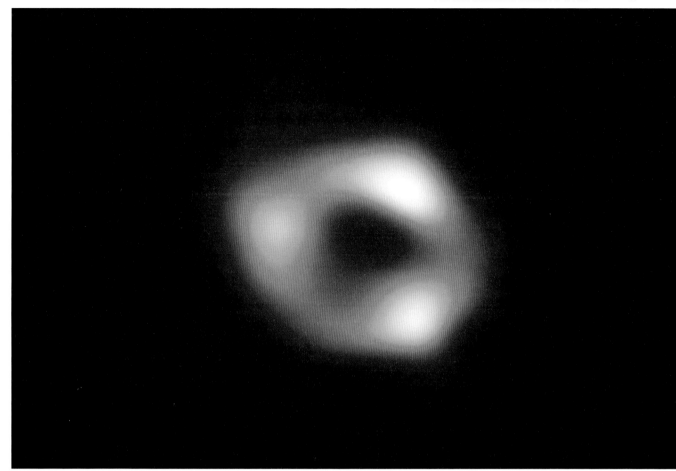

ABOVE The first image of the supermassive black hole at the centre of our own Milky Way galaxy, produced by the Event Horizon Telescope in 2022. The black hole is about 27,000 light years away from Earth.

Hawking now realized he would need to investigate the empty space at and near the black hole's event horizon from the quantum perspective. "He mulled over the Zel'dovich/Starobinsky discovery for several months," recalled Thorne, "looking at it first from one direction and then from another".

Then came the breakthrough. According to Hawking's daughter, Lucy, the significance of a quantum perspective became clear during a warm-up mathematical exercise. As he did his calculations, he found to his "surprise and annoyance" that his results suggested the opposite of what he'd hoped. By December 1973 he'd realized that not only did black holes – even when they were not spinning – radiate heat, but also that they did so by the amount required if the area of their event horizons was indeed a measure of their entropy. This marked a milestone for "black hole thermodynamics", and led to Hawking's insight that black holes are not so black: they give off what would become known as "Hawking radiation".

BLACK HOLE EXPLOSIONS
By early 1974, Hawking had developed this work into a fully-fledged theory, and on 17 January he submitted a paper to the journal *Nature*, which was published in March. He gave it the provocative title: "Black hole explosions?" and it began, "Quantum gravitational effects are usually ignored in calculations of the formation and evolution of black holes." He went on to explain that "as a black hole emits this thermal radiation, one would expect it to lose mass. This in turn would increase the surface gravity and so increase the rate of emission ... For a black hole of solar mass this is much longer than the age of the Universe." In other words, a black hole formed from a mass equivalent to our Sun would take an eternity to evaporate. However, there might, he pointed out, be much smaller black holes which were formed by fluctuations in the early universe. Near the end of its life the rate of emission from such a primordial black hole would be very high, Hawking wrote, representing "a fairly small

explosion by astronomical standards but equivalent to about 1 million 1 Mton [megaton] hydrogen bombs".

At the time, new satellites with gamma-ray detectors were being sent to space, which would be able to detect such explosions, and he wrote his paper in the hope of quick observational confirmation (which would have led to a Nobel Prize). Alas, even after half a century there is still no definite evidence for either evaporating or non-evaporating primordial black holes. The theory is not wrong, however: it is just that these primordial black holes are now thought to be rare.

The reason black holes "ain't so black", as Hawking put it in *A Brief History of Time*, is ultimately down to one strange consequence of quantum theory: that empty space is not empty at all.

Quantum mechanics says that when you examine something there is always some intrinsic uncertainty as to what you are looking at. This is even true for empty space itself: when you look closely enough, what you thought was empty space is alive with particle antiparticle pairs constantly popping in and out of existence. Usually the pairs would recombine and cancel out, but if they appear on the border of the event horizon of a black hole – the point of no return – they may find themselves on adjacent sides, with one sucked in, and the other released. When the latter particle escapes, the black hole loses a small amount of its energy, and therefore some of its mass (mass and energy are of course related by Einstein's equation $E = mc^2$). The result is that the radiation gradually takes mass away from the black hole – it evaporates. In effect, a black hole rots from the outside in. Consequently, because it is evaporating, a black hole will have a finite lifespan. It is this discovery we now call "Hawking radiation", and it is Hawking's greatest achievement.

ROLLING CANDY ON THE TONGUE

The prediction is such a beautiful result – one that unifies three great edifices of physics, quantum theory, general relativity, and thermodynamics – that most if not all experts accept that it is correct. John Wheeler once said that just talking about it was like "rolling candy on the tongue".

Hawking's discovery half a century ago stunned his peers. Testament to its importance is that it still guides the ongoing search for a more complete theory of quantum gravity. Hawking radiation is the one result in the continuing attempt to reconcile quantum mechanics and gravity that is accepted by the entire community of physicists working on the subject.

It was when I attended Hawking's sixtieth birthday celebrations in Cambridge that he declared that "I would like this simple formula to be on my tombstone." Hawking's famous equation of the entropy of a black hole does indeed adorn his gravestone.

SEE ALSO:
————

The prediction is such a beautiful result – one that unifies three great edifices of physics, quantum theory, general relativity, and thermodynamics ... John Wheeler once said that just talking about it was like "rolling candy on the tongue".

In Galileo's Orbit

Galileo Galilei was one of Stephen Hawking's major inspirations. The reason can be found at the end of *A Brief History of Time*. Though only two of its 198 pages are dedicated to the seventeenth-century astronomer and thinker, even that is perhaps generous given the book's subject matter: while the universe has been around for 13.7 billion years, Galileo lived for only 77. However, "Galileo", Hawking declares, "perhaps more than any other single person, was responsible for the birth of modern science." A bold claim, and one that is hard to verify.

In Hawking's office was a paperback translation of Galileo's most famous work, *Dialogues Concerning Two New Sciences*. Originally published in Italian, the book was dedicated to Galileo's powerful patron, Ferdinando II de' Medici, the Grand Duke of Tuscany, who received the first printed copy on 22 February 1632. In the Dialogue, various questions are discussed and debated in a witty conversation between three men: Simplicio, a philosopher who defends the old views of Aristotle; Sagredo, an intelligent layman; and Salviati, who represents Galileo himself.

HURRIED LITTLE MASTERPIECE
In *On the Shoulders of Giants*, Stephen Hawking's meditation on the great works of science, he describes how Salviati and Sagredo put forward "persuasive" Copernican arguments, while Simplicio defended Aristotle and Ptolemy's traditional notion that the Earth was orbited by the Sun. In other words, Galileo backed the Copernican view that the Earth and other planets orbit the Sun. Hawking's brief biography of him in *A Short History of Time* relates how this "annoyed the Aristotelian professors", the Church declared Copernicanism "false and erroneous", and the Pope "brought Galileo before the Inquisition, who sentenced him to house arrest for life".

Also in Hawking's office was a copy of the major biography of Galileo by the distinguished scholar of the Scientific Revolution, John Heilbron of the University of California, Berkeley. It had been published in 2010, 400 years after Galileo's treatise *Siderius nuncius*, "Starry Messenger", a "hurried little masterpiece", as Heilbron put it, that had recorded Galileo's remarkable observations using the recently invented telescope.

Galileo's own astronomical discoveries had only increased his conviction that the Earth moved around the Sun. As one example, he observed that Venus exhibits a full set of phases, similar to those of the Moon. That was significant support for the Copernican view, because Ptolemy's geocentric model predicted that Venus's disk would show only the Moon in its new (i.e. dark) and crescent phases as it orbits Earth.

On reading what little Hawking says about Galileo in *A Brief History of Time*, Heilbron is struck by what he calls an "interesting epistemological bias" that glosses over the personal and religious context of Galileo's life, no doubt in accordance with the agenda of Hawking's book. For example, Hawking says that Galileo's conflict with the

ABOVE Galileo Galilei, 1852. Found in the collection of the Russian State Library, Moscow.

Dialogues Concerning
Two New Sciences
Galileo Galilei

Translated by Henry Crew and Alfonso de Salvio

Galileo did not keep his side of the bargain.

Church was "central to his philosophy". "In a way that's true," says Heilbron, but it was not, as Hawking wrote, Galileo's assertion that humankind could understand the world that riled the Church: rather Galileo's challenge to its authority in the realm of natural science. That is why it was not the Aristotelian professors who initiated the persecution of Galileo (this was also Hawking's opinion) but the Dominicans who oversaw the censorship system and condemned his teaching as contrary to scripture. "Again," said Heilbron, "Stephen makes it solely an academic argument, neglecting personalities and institutions." At the pinnacle of the censorship system was Pope Urban VIII. A friend, admirer and patron of Galileo, he was willing to allow Galileo to publish what became *The Dialogue Concerning the Two Chief World Systems*, so long as Galileo did not take sides in the debate about Copernican theory and concluded that the truth in astronomy lay beyond the grasp of unaided human reason.

GALILEO'S PUNISHMENT

Galileo did not keep his side of the bargain, however. By giving Simplicio the final word, that God could have made the universe any way he wanted and look any way he wanted, Galileo put one of Urban's favourite arguments in the mouth of the person who had lost the debate.

One manifestation of the Pope's dismay was Galileo's punishment. Rather than being sent to a monastery to be rehabilitated and released after a year or two, which was the standard punishment for someone convicted, like Galileo, of raising "a strong suspicion of heresy", Galileo was put under house arrest. This was perhaps better than monastic isolation, but he would never be granted his freedom by Urban. "It was quite personal at the end," says Heilbron.

More than three and a half centuries later, following a long investigation by a panel of scientists, theologians and historians, the Catholic Church itself partially recanted. In 1984 its preliminary report said that Galileo had been wrongfully condemned. Then, at the end of October 1992, Pope John Paul II, addressing the Pontifical Academy of Sciences, confronted the Galileo affair:

There exist two realms of knowledge, one which has its source in Revelation and one which reason can discover by its own power. To the latter belong especially the experimental sciences and philosophy. The distinction between the two realms of knowledge ought not to be understood as opposition. The two realms are not altogether foreign to each other; they have points of contact. The methodologies proper to each make it possible to bring out different aspects of reality.

But the Pope went on to say that Galileo failed to make a distinction between the scientific approach to natural phenomena and more general reflections on the natural world of a philosophical order, which a scientific approach generally requires. Many commentators regard the suggestion that the Church and Galileo should somehow share the guilt for the whole episode – as Heilbron puts it, that "Galileo was right about interpretation of scripture and the theologians right about the true nature of science" – as "a bizarre and unhelpful compromise".

SECULAR DEITY

It is ironic that, over the centuries following his death, Galileo himself evolved into something of a secular deity, with monuments erected to him and even bones from his hand displayed as relics (they are the most popular exhibit at the Museo Galileo in Florence). Stephen Hawking could never resist mentioning that he was born on the three-hundredth anniversary of Galileo's death, and neither could the media, as though this fact might somehow help us comprehend the origins of his genius. In a similar vein, some of Galileo's followers liked to point out that Galileo was born in 1564, the year in which Michelangelo died. "I don't know whose bodies the spirit inhabited between Michaelangelo and Stephen Hawking," reflects John Heilbron, "but one of them was surely Newton's, who was born, more or less (it depends on whether the necromancer uses the Julian, Gregorian, or both calendars) in the year of Galileo's death."

SEE ALSO:

Hawking and God, p.54
On the Shoulders of Lucasian Giants, p.62

OPPOSITE Galileo's most influential scientific work, *Dialogues Concerning Two New Sciences*, translated by Henry Crew and Alfonso de Salvio and published in 1954.

"If we discover a complete theory [of the universe], it would be the ultimate triumph of reason – for then we should know the mind of God."

Hawking and God

At the end of *A Brief History of Time*, Stephen Hawking makes a much-quoted reference to a deity: "If we discover a complete theory [of the universe], it would be the ultimate triumph of reason – for then we should know the mind of God." He was not using this mention of God in a religious way, but as a metaphor for the laws of nature.

Even so, it seems somehow fitting, given the cosmic ambition of his science, that Stephen Hawking was a member of the Pontifical Academy of Sciences, as shown by this medal. Established in 1936 by Pope Pius XI in the Casina Pio IV, in a beautiful villa in the heart of the Vatican Gardens in Rome, the Academy aimed to promote the progress of mathematical, physical and natural sciences, along with the study of related philosophical problems.

ANXIOUS FOR KNOWLEDGE

The Vatican's interest in science dates back centuries earlier, to the Accademia dei Lincei (Academy of the Lynxes) which was founded in Rome in 1603 as the first exclusively scientific academy in the world, with the motto Sapientiae Cupidi – "Anxious for knowledge". Science was closer to the Church at that time, as priests studied the world as, they believed, God's creation.

The relationship between the Catholic Church and science has changed considerably in the three centuries between 1603 and Hawking. When it came to the Academy, Hawking did not have to be interested in religion. The membership, made up of 80 of the world's leading scientists, is based on merit rather than beliefs and is not restricted to those professing the Christian faith. "It has members of all faiths and none," said his friend Martin Rees. Conceived as a place where science and faith could meet in discussions of the biggest ideas, the Academy provides a fascinating lens through which to examine Hawking's views.

His relationship with the Academy dates back to the Seventies, when he received the Pius XI Medal, first awarded in 1962 and, explained Lord Rees, given every two years to a promising young scientist under the age of 45.

Hawking received the gold medal in 1975, when he was 33, just after doing his ground-breaking work on Hawking radiation, which shows that the Vatican was as impressed by his research as his peers. He even presented to the Academy a summary of his research on black holes.

THE NO-BOUNDARY PROPOSAL

At another conference in the Vatican, in 1981, Stephen Hawking had first put forward the suggestion that time and space together formed a surface that was finite in size, but had no boundary or edge. It was a precursor of the "no-boundary proposal" he was developing with his long-term collaborator, James Hartle from the University of California, Santa Barbara.

In the elegant villa that housed the Academy, Hawking outlined what he would later regard as his most important idea: that the universe could have arisen from nothing. "There ought to be something very special about the boundary conditions of the universe," he told the Academy, "and what can be more special than the condition that there is no boundary?"

Over the years, journalists in search of a Hawking headline have sought the assistance of God. In 2010, one British newspaper, for its serialization of his new book, ran a front page trumpeting, "Hawking: God did not create universe", as though this was a scoop. In *The Grand Design*, written with Leonard Mlodinow, Stephen Hawking had indeed described how M-theory (a possible ultimate theory of the universe, uniting general relativity and quantum mechanics) may offer answers to the question of creation. "According to M-theory, ours is not the only universe," wrote Hawking. "Instead, M-theory predicts that a great many universes were created out of nothing. Their creation does not require the intervention of some supernatural being or God."

THE MIND OF GOD

Stephen Hawking's line in *A Brief History of Time* about the ultimate triumph of human reason is perhaps his most famous quotation of all and, in the subsequent newspaper article, what was referred to as Stephen Hawking's "mind of God" pronouncement was billed as his "previous view" on religion. But James Hartle himself told me that he could not recall a single discussion with Stephen about God.

OPPOSITE Membership insignia (medal) of the Pontifical Academy of Sciences, a scientific academy in Vatican City which Hawking had participated in since the 1970s and became a member of in 1986.

Even so, over the years, journalists have tried to make too much of Stephen's Hawking's apparently religious pronouncements: as he told the Spanish publication *El Mundo*, "What I meant by 'we would know the mind of God' is, we would know everything that God would know, if there were a God, which there isn't. I'm an atheist."

PLAYFUL LOCUTIONS

Juan-Andres Leon, curator, thinks that Hawking's playful locutions on God and creation are likely to have boosted publicity and sales for his book and that, once he became famous, editors may have regarded such ambiguity as a definite asset. For example, later editions of *A Brief History of Time* removed the preface by Carl Sagan with its decidedly more atheistic tone. Leon adds that Stephen's relationship with the Vatican

was a beneficial scientific alliance, having started at a time when the church opened up to modern theories, including Darwin's evolution and the Big Bang; encounters at the Academy allowed Hawking to meet influential scientists from around the world, and to become one of them, long before he was a celebrity.

In 2001, when I interviewed Stephen Hawking, he made an illuminating comment on where he stood in relation to God and science. "If you believe in science like I do", he told me, "you believe that there are certain laws that are always obeyed. If you like, you can say the laws are the work of God, but that is more a definition of God than a proof of his existence." In a *Daily Telegraph* piece of his I edited in February 2008, he described how, at the 1981 conference on cosmology at the Vatican, the Pope had even told the delegates that they should not inquire into the beginning of the universe itself, because that was the moment of creation and the work of God. "I was glad he didn't realize I had already presented a paper at the conference investigating precisely that issue," joked Hawking. "I didn't fancy the thought of being handed over to the Inquisition like Galileo."

ULTIMATE ENVIRONMENTAL SCIENCE

Three-and-a-half decades later, Hawking's view of creation continued to differ from the Church's, as can be seen from his contribution to the Academy's Plenary Session of 2016 on "Science and Sustainability". Cosmology, says Lord Rees, is the ultimate environmental science and, in a talk on "The Origin of the Universe", Hawking must have delighted the Vatican with his reference to the work of Georges Lemaître, a Belgian Catholic priest and theoretical physicist at the Catholic University of Louvain who had served as President of the Pontifical Academy of Sciences from 1960 until his death in 1966. Lemaître's work in the first half of the twentieth century developed what we now call the "Big Bang theory" of the beginning of the universe; a theory grounded in physics, which also sat with his belief in God as creator.

After pointing out in his talk how Edwin Hubble's discovery of the expansion of the universe had transformed the debate about whether the universe had a beginning – "if one traces the motion of the galaxies back in time, they would all have been on top of each other about 14 billion years ago" – Hawking added that Lemaître was the first to propose a model in which the universe had such an infinitely dense beginning. But though Hawking and Lemaître agreed on a beginning of the universe – Hawking's no-boundary proposal "describes how our familiar notions of space and time can come into being, thus realizing Georges Lemaître's vision" – they disagreed on whether a creator was needed to start this process. Unlike Lemaître, Hawking did not see a role for religion or God:

This would remove the age-old objection to the universe having a beginning, that it would be a place where the normal laws broke down. The beginning of the universe would be governed by the laws of science.

WHY DOES THE UNIVERSE BOTHER TO EXIST?

In 1992, when he appeared on the long-running radio show *Desert Island Discs*, he discussed his no-boundary model, under which the universe has neither beginning

In the elegant villa that housed the Academy, Hawking outlined what he would later regard as his most important idea: that the universe could have arisen from nothing.

nor end, but emphasized that it did not mean the way the universe began "was a personal whim of God". "You still have the question," he added: "why does the universe bother to exist? If you like, you can define God to be the answer to that question." This view chimes with that of Einstein, who famously declared, "I want to know how God created the world", but was careful to explain that he was talking about Spinoza's God, by which Einstein meant one "who reveals himself in the harmony of all that exists, not in a God who concerns himself with the fate and the doings of mankind".

Stephen Hawking's discussions at the Vatican were not confined to cosmology. At the end of 2016, he contributed to a workshop on the power and limits of artificial intelligence (AI) and warned: "AI will be either the best or the worst thing ever to happen to humanity. We do not yet know which." He was one of the co-signatories to a symposium statement on the rise of intelligent machines that declared: "All governments should be alerted that a major industrial revolution is underway and must take new measures to manage it."

The year before he died, Hawking also signed a declaration on climate change and air pollution, organized by the Academy, which showed that he was well aware humanity was entering a dangerous phase of the Earth's history. "The very fabric of life on Earth, including that of humans, is at grave risk," ran the declaration. "We propose scalable solutions to avoid such catastrophic changes. There is less than a decade to put these solutions in place to preserve our quality of life for generations to come. The time to act is now."

SEE ALSO:
———————
A World of Souvenirs, p.200

ABOVE Group photograph with Pope Francis and Stephen Hawking at the Vatican's Pontifical Academy of Sciences in 2016.

THE
VERY
EARLY
UNIVERSE
Edited by G·W·Gibbons
S·W·Hawking
S·T·C·Siklos

Workshop on the Baby Cosmos

In the middle of Stephen Hawking's bookshelves, above his microwave oven, is a thick volume with a striking red, white and blue graphic of cosmic creation on the cover. Inside is a record of one of the most important gatherings of modern cosmology, and it was Hawking who helped bring it about, convening the best people on the planet to work together on the hottest topic in the field. Drily entitled *The Very Early Universe*, but subsequently lauded by other scientists, the book contains papers from 29 different authors who attended a 1982 workshop held in Cambridge between 21 June and 9 July to discuss "recent advances in understanding of the origin of the universe".

The volume was edited by Hawking himself along with his Cambridge colleagues Stephen Siklos, a former student, and Gary Gibbons. Gibbons remembers how – unusually – the workshop brought together scientists from the two sides of the Iron Curtain, notably from the US and the USSR. Cunningly, they invited senior Soviet academicians to bring along their assistants, a way of securing the attendance of rising stars like Andrei Linde and Alexei Starobinsky, who were their real target.

INFLATION
It followed the earlier workshop, also funded by the Nuffield Foundation and hosted by the University of Cambridge, that had led to the graffiti-strewn blackboard. This one, however, explains Gibbons, was devoted to models of the universe's birth, when it was less than a second old, the Big Bang model and the then brand-new theory of inflation. "This theory took off very quickly."

While the Big Bang model suggests that the universe expanded into its current state from a primordial pinpoint of enormous density and temperature, the theory of inflation embellishes this idea by proposing that at the very beginning – at around 10 to the power of -35 seconds after the Big Bang – the rate of this expansion was exponential, so the universe ballooned by a factor of around 10 to the power of 60 – that's a 1 with 60 zeros at the end – within a tiny fraction of a second.

PRIMORDIAL PERTURBATIONS
Aside from putting the basic mechanics of inflation on a firmer footing, the meeting was distinguished by something else: the delegates showed how "primordial perturbations" in the baby universe paved the way for the galaxies and other huge structures dotted around us in the cosmos. "This meeting solved various puzzles," said Gibbons,

such as the lack of primordial 'fossils' (primordial black holes and magnetic monopoles, elementary particles with only one magnetic pole) in the early universe. But importantly, it also led to the realization that the quantum fluctuations should be reflected in fluctuations or deviations of temperature of the cosmic microwave background, which are centre stage of current work.

OPPOSITE Proceedings of the 1982 Nuffield workshop *The Very Early Universe*, published by Cambridge University Press the following year. This workshop was considered crucial in the development of cosmic inflation, the theory that describes the very rapid expansion of the early universe.

Gibbons remembers how – unusually – the workshop brought together scientists from the two sides of the Iron Curtain.

By the end of the workshop, a new paradigm had emerged, which has been changed little since.

In other words, this second Nuffield workshop was, if anything, more influential than the first, and is the reason why Stephen Hawking kept an inflatable beach ball in his office. The ball was decorated with a picture of the baby universe, the afterglow of the Big Bang, based on data gathered two decades later by NASA's Wilkinson Microwave Anisotropy Probe. The images from the probe vindicated the insights gained at the 1982 meeting much more convincingly than the lower-resolution data from an earlier NASA probe, called COBE, which in the 1990s Hawking had rashly hailed as "the most important discovery of the century, if not of all time".

This 1982 Nuffield meeting worked out the details of how quantum fluctuations in the early universe could lead to large-scale structure in the universe: quantum theory predicts that fluctuations in a field that causes inflation arise on very tiny scales, too small to create the lumpiness needed to seed galaxies. That problem is fixed by the tremendous expansion during inflation that blows these fluctuations up from subatomic to astrophysical size.

A NEW PARADIGM

With a commitment of only a couple of seminars each day at the workshop, there was a great deal of opportunity to do research. The details of inflation, a concept put forward by Alan Guth while at Stanford University, and the quantum origin of density fluctuations, were worked out in detail by "quite a remarkable group of people", according to Michael Turner of the University of Chicago, one of those present. Problems remained with Guth's concept, but over three weeks the implications of inflation for the large-scale structure of the universe were thrashed out by four groups working separately. Stephen Hawking would invite delegates home to dinner, or to croquet matches, as a respite from their calculations. "That area was very competitive and hard working," says Gibbons, adding that the calculations did not require computers.

"There was tension in the air because we all realized that inflation could make the seeds for galaxies, but what was really exciting was that people initially got different answers," recalled Turner. He had investigated one approach with Paul Steinhardt, but their calculations disagreed with Hawking's and Starobinsky's. Guth, meanwhile, was still working out his own result until the very end of the meeting. The problems with his original concept seem to have been ironed out in a new version called "slow-roll inflation" developed by Andrei Linde and, independently, by Andreas Albrecht and Paul Steinhardt.

Linde's fix was brilliant, said Turner, though many of the details were wrong and still needed to be filled in. By the end of the workshop, the finer details of slow-roll inflation were worked out, and all four groups agreed on how quantum fluctuations gave rise to the seeds for galaxies. A new paradigm had emerged, which has been changed little since.

"That was a really famous meeting," says Stephen Hawking's biographer, Graham Farmelo. "This gathering represents a remarkable moment, where the inflation theory really matured."

Two decades later there was another conference/workshop on the same topic, gathering together 15 of the original 30 participants. The 2007 meeting was attended by Guth, as well as Linde, now from the Department of Physics at Stanford University, along with Steinhardt and Stephen Hawking. Guth opened the conference by pointing out that while there were 11 talks on the opening day in 2007, back in 1982 there had been only nine during the whole of the first week. "That left lots of time for speaking with one another, working or just thinking!" recalled Michael Turner. "And, of course, the field of early-universe cosmology was much smaller then."

"I KNOW HOW THE UNIVERSE BEGAN"

At the later meeting, Andrei Linde recalled how late one night in 1982 he had woken up his wife, the physicist Renata Kallosh, to tell her, "I know how the universe began." That, he said, launched "the most exciting field of science in the last twenty-five years". Turner, who gave the concluding talk and wrote up the meeting for the journal *Nature*, said that the 1982 meeting had changed the course of cosmology, and that the 2007 reunion was mostly to "get the band back together" to celebrate the fact. "The Nuffield meeting was the most exciting I have ever been to."

SEE ALSO:

Blackboard of Scientific Doodles, p.80
Universe on a Beach Ball, p.150

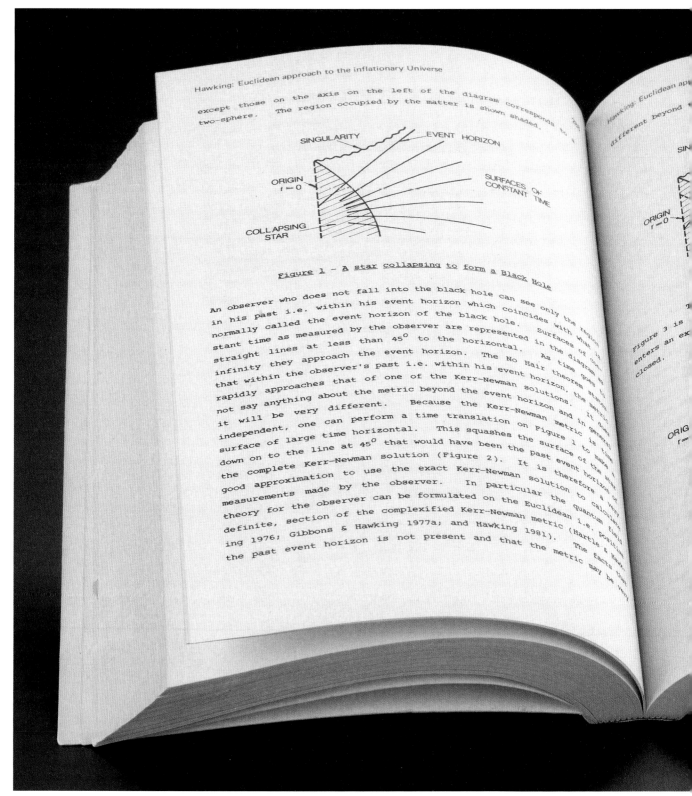

Hawking: Euclidean approach to the inflationary Universe

except those on the axis on the left of the diagram corresponds to a two-sphere. The region occupied by the matter is shown shaded.

SINGULARITY — EVENT HORIZON

ORIGIN
r = 0

SURFACES OF
CONSTANT TIME

COLLAPSING
STAR

Figure 1 – A star collapsing to form a Black Hole

An observer who does not fall into the black hole can see only the region in his past i.e. within his event horizon which coincides with what is normally called the event horizon of the black hole. Surfaces of constant time as measured by the observer are represented in the diagram by straight lines at less than 45° to the horizontal. As time goes to infinity they approach the event horizon. The No Hair theorem states that within the observer's past i.e. within his event horizon, the metric rapidly approaches that of one of the Kerr–Newman solutions. It does not say anything about the metric beyond the event horizon and in general it will be very different. Because the Kerr–Newman metric is time independent, one can perform a time translation on Figure 1 to make the surface of large time horizontal. This squashes the surface of the past down on to the line at 45° that would have been the past event horizon of the complete Kerr–Newman solution (Figure 2). It is therefore a very good approximation to use the exact Kerr–Newman solution to calculate measurements made by the observer. In particular the quantum field theory for the observer can be formulated on the Euclidean i.e. positive definite, section of the complexified Kerr–Newman metric (Hartle & Hawking 1976; Gibbons & Hawking 1977a; and Hawking 1981). The facts that the past event horizon is not present and that the metric may be very

Hawking: Euclidean ap

different beyond

SIN

ORIGIN
r = 0

Figure 3 is

enters an ex

closed.

ORIG
r =

ABOVE A page from the Proceedings of the 1982 Nuffield workshop *The Very Early Universe*, showing Hawking's entry on "Euclidean approach to the inflationary Universe".

On the Shoulders of Lucasian Giants

Stephen Hawking held the Lucasian Chair of Mathematics, which had previously been occupied by some of the scientific greats, notably Isaac Newton and Paul Dirac. He was in awe of them yet he did more to make the chair famous than any of his illustrious predecessors.

The post of Lucasian Professor is an ancient one, founded in 1663 by Henry Lucas, Cambridge University's Member of Parliament from 1639 to 1640. Lucas bequeathed land to the university sufficient at the time to provide an income of £100 a year to fund a professorship.

The inaugural Lucasian professor was Isaac Barrow, a Christian theologian and mathematician. Other notable holders include Charles Babbage, a polymath who came up with a programmable computer, who held the chair from 1828 to 1839 and whose archive and mechanical calculators are held by the Science Museum Group, and Michael Green, a pioneer of string theory, who was in post from 2009 to 2015.

Of all Hawking's Lucasian predecessors, however, there were two towards whom he felt a particular swell of pride. As he himself put it, "I must say I like the feeling that I hold the same job as Newton and Dirac."

PRINCIPIA
Isaac Newton was Barrow's student, and among the most influential scientists of all time. During the plague years of 1665 and 1666 he made huge strides in mathematics, optics and celestial dynamics, revealing for the first time Nature's unity through scientific laws that ruled the heavens and the Earth. The presentation of Part One of his *Philosophiæ Naturalis Principia Mathematica* (usually referred to as the *Principia*) to the Royal Society on 28 April 1686 marked a turning point in science. Some rate *Principia* as the greatest scientific book of all time; others liken it to a great cathedral of ideas soaring above the ramshackle constructions around it. To Hawking it was "the most influential book ever written in physics".

Newton put forward three laws of motion that rewrote the science of moving objects and showed that bodies on Earth and in the heavens were governed by one and the same force: gravity. His law of universal gravitation is perhaps the most famous of all, stating that the gravitational tug between two bodies is proportional to the masses of the bodies and diminishes with distance as the square of their separation.

By uniting terrestrial and celestial mechanics, Newton also solved a problem that had obsessed humankind since the dawn of consciousness: the movement of planets, which provided the key to timekeeping and navigation.

As the greatest of the Augustan poets, Alexander Pope, wrote in his celebrated epitaph on Newton's tomb in Westminster Abbey:

Nature and Nature's Laws lay hid in Night.
God said, Let Newton be! and All was Light.

Even today, the trajectories of moons, missiles, rockets, satellites and space probes such as Voyager 1 – the most distant human-made object from Earth – are calculated in advance using Newton's centuries-old theory.

NOT A PLEASANT MAN
To be able to make predictions from his laws about the motion of planets and probes, Newton discovered the branch of mathematics known as calculus, which remains a cornerstone of mathematics and theoretical physics. However, his poor behaviour during a "priority dispute" with the German polymath Gottfried Wilhelm Leibniz, over who could justly claim to have invented calculus (even though elements had been glimpsed long before in ancient Greece, Egypt, India and the Middle East), was one of the reasons Hawking could remark in *A Brief History of Time* that "Isaac Newton was not a pleasant man."

Newton's theory of gravity would, however, eventually be overturned centuries later by another of Hawking's heroes, Albert Einstein. His general theory of relativity saw gravity not as a force but as warped space-time. As

OPPOSITE English translation of Principia Mathematica in which Isaac Newton described gravity and the laws of planetary motion, along with a method of inquiry that became the gold standard for science.

Sir Isaac Newton
PRINCIPIA
VOL. II THE SYSTEM OF THE WORLD

Motte's Translation
Revised by Cajori

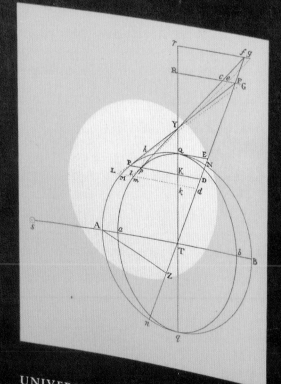

UNIVERSITY OF CALIFORNIA PRESS

By uniting terrestrial and celestial mechanics, Newton also solved a problem that had obsessed humankind since the dawn of consciousness: the movement of planets, which provided the key to timekeeping and navigation.

TOP RIGHT Sir Isaac Newton. **ABOVE** English physicist Paul Dirac. Both Newton and Dirac held the Lucasian chair before Hawking.

one physicist put it, "Space-time grips mass, telling it how to move ... Mass grips space-time, telling it how to curve."

ANTIMATTER

In 1933, when Einstein became the first staff member of the Institute for Advanced Study in Princeton, New Jersey, he was asked who he wanted to join him. Top of his list was an English physicist, Paul Dirac, who had taken up the Lucasian chair in 1932, around the same age that Newton had been given the prestigious role. A few months later, at the age of 31, Dirac became the youngest theoretician to win the Nobel Prize for Physics. "Dirac knew that the Chair was more than an accolade," says Graham Farmelo, biographer of both Dirac and Hawking: "it was a vote of confidence but also a challenge. He was expected to ... leave a legacy that scientists would talk about for centuries. By no means all the holders of the Lucasian Chair had justified their promise: William Whiston, John Colson and Isaac Milner are in no one's list of great mathematicians or scientists."

Although he remains relatively unknown to the public, in the 1920s and 1930s Dirac was a pioneer of quantum mechanics. His wave equation for describing the electron was published in 1928. Now known as the Dirac equation, it was a marriage of special relativity and quantum theory, a milestone on the quest to unify the laws of physics that would later consume Hawking himself.

The equation only worked, however, if one assumed that there was such a thing as an "anti-electron". Here Dirac was dramatically proved right: the anti-electron (positron) was discovered experimentally by the American Carl Anderson in August 1932, and shortly afterwards the concept of "antimatter" became a cornerstone of physics.

"Probably the greatest British theoretical physicist since Newton", is how Stephen Hawking described Paul Dirac at his Westminster Abbey commemoration service in November 1995, where Hawking also lamented that it had taken too long to recognize the genius of the inventor of antimatter, who made so many other fundamental contributions, as articulated in his milestone work, *The Principles of Quantum Mechanics* (1930).

Because of these remarkable antecedents, Hawking himself did not initially feel worthy of the role of Lucasian professor. He did not apply for the job, he told Hélène Mialet, now at York University, Canada, who was writing a history of the Lucasian chair. After all, he explained, he was already a professor, and hoped that the university would "bring in someone good" from outside Cambridge, such as Sir Michael Atiyah, who had won one of the great mathematics awards, the Fields Medal, and would go on to be President of the Royal Society. "I was rather disappointed when George Batchelor, who was then head of DAMTP, told me the electors had chosen me," said Hawking. "I urged them to think again and get someone like Atiyah. But Batchelor was very much against the idea of giving the Lucasian professor to Atiyah, whom he regarded as a pure mathematician thinly disguised as a physicist ... In the end I agreed to accept if I was given certain support."

Stephen Hawking became the Lucasian Professor of Mathematics in 1979, but still regarded himself as "a stopgap" because "they thought I wouldn't live very long." As it turned out, he lived with motor neurone disease for another 39 years.

LITTLE SMATTERERS

In one sense, Hawking was quite different from his distinguished predecessors. Newton thought that matters of truth, whether natural philosophical or theological, should be handled only by experts, not the hoi polloi and "little smatterers", and deliberately made his *Principia* obscure. Dirac had no interest in publicizing or popularizing his work either, feeling it could only be expressed in the form of mathematics and, to be specific, beautiful equations. "To draw its picture is like a blind man touching a snowflake," he said. "One touch and it's gone."

Hawking, on the other hand, loved to be noticed, and helped make the chair prominent too. In "All Good Things...", the finale of *Star Trek: The Next Generation* that aired in May 1994, the android Data has, in a parallel universe (of course), become Lucasian Professor and even possesses a copy of Hawking's *A Brief History of Time* in his library. "Holding the Lucasian chair does have its perquisites," Professor Data explains to Captain Jean-Luc Picard: "This house originally belonged to Sir Isaac Newton when he held the position."

In the space of just 15 years, Hawking had made the centuries-old Lucasian chair famous to the point that it appeared on one of the most recognizable media franchises of all time.

SEE ALSO:

Young Hawking, p.28
Royal Recognition, p.102
To Boldly Go..., p.126
Universe on a Beach Ball, p.150
Stephen Hawking, You Are Wrong!, p.176
Hawking's Equation, p.228

Not the New Einstein?

In Stephen Hawking's office was a framed facsimile of a handwritten manuscript by Albert Einstein that contained his most famous equation – perhaps the most famous of all – and explained his special theory of relativity to the lay reader.

Hawking was only too pleased to accept the facsimile, provided by the Hebrew University of Jerusalem and collected by the Einstein Papers Project: he had profiled Albert Einstein along with Galileo and Newton at the end of *A Brief History of Time*, and such was his respect for Einstein that he would dismiss claims he was "the new Einstein" as "media hype". However, "it is ironic that Einstein was one of Hawking's heroes," commented Jürgen Renn, Director at the Max Planck Institute for the History of Science in Berlin, "since he was sceptical about the reality of black holes – Einstein saw it as an artefact of the mathematics, a border beyond which his theory of general relativity would no longer be valid."

Einstein's long-hand manuscript, which is one of only three documents in the world containing the famed formula in his own handwriting, was for an article entitled "E=mc²: On the Most Urgent Problem of Our Time" published in the magazine *Science Illustrated*. In the article, he warned how this little equation brings with it a great threat of evil: "Averting that threat has become the most urgent problem of our time."

The equation itself is easy to understand: E (energy) is equal to m (mass) multiplied by c (the speed of light) multiplied by itself. The speed of light expressed in metres per second (approximately 300,000,000, which is 186,000 miles per second), gives a huge number when squared: a 9 followed by 16 zeros: a tiny amount of mass can be converted into a colossal amount of energy, as the first nuclear weapons detonated in Hiroshima and Nagasaki in 1945 had revealed in a horrifying way.

The formula dates to Einstein's annus mirabilis in 1905 when, at the age of just 26, he published papers which tore up established Newtonian theory and laid the foundations of modern physics theory. A few years later, one of those papers had matured into his general theory of relativity, by which time Einstein had helped to redraw our understanding of the nature of matter, energy, motion, space and time, laying the foundations for Hawking's work in the future.

EINSTEIN'S "COBBLER'S JOB"

Einstein had graduated from the Swiss Federal Polytechnic in Zurich, or ETH, only five years earlier, with good, but not great, grades. His attempts to become an entry level physics lecturer came to nothing – his professors thought him confrontational, and he could indeed be acerbic and arrogant. But thanks to an old friend, he managed to get a job at the Patent Office in Berne. Though a modest town of 60,000 inhabitants, it was the capital of the Swiss Confederation and an intellectual powerhouse, the seat of the International Postal Union, International Office of Railway Transport, International Telegraphic Union and the administration of the Paris Convention, an intellectual property treaty. And, of course, there was the Patent Office.

Some claim that what Einstein called his "cobbler's job" was the key to his astonishing intellectual outburst in 1905. By one view, it was having to appraise inventions that helped hone his creativity. By another, Einstein was inspired by patents on timekeeping: in the new age of mass communications innovations like the telegraph and the steam train required accurate time-keeping, so the concept of time was likely to have played on his mind.

OPPOSITE Poster of Isle of Man first day covers of postage stamps celebrating 100 years of General Relativity, Einstein's theory of gravity.

Such was Hawking's respect for Einstein that he would dismiss claims he was "the new Einstein" as "media hype".

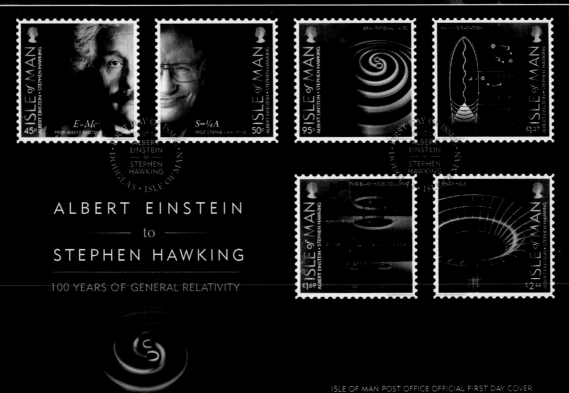

THIS PAGE Framed facsimiles of Albert Einstein's manuscript, featuring the handwritten version of the most famous formula in physics, E=MC2, presented to Stephen Hawking by The Hebrew University of Jerusalem, 2006. The article appeared in print in 1946 in the magazine *Science Illustrated*.

gesträubt hatte, eine solche „Erzeugung" von Wärme zuzugeben, gelang es schliesslich zu zeigen, dass zur Erzeugung einer Wärmeenergie durch Reibung immer ein genau proportionaler Betrag von mechanischer Energie aufgewendet werden muss). Dadurch wurden die Sätze von der Erhaltung der mechanischen und der thermischen Energie zu einem einzigen Satz verschmolzen. Hier gelangt bedrängte sich den Physikern die Überzeugung auf, dass dies Erhaltungsgesetz auch auf die chemischen und elektromagnetischen Vorgänge auszudehnen sei, welche Erweiterung sich bisher auf allen Gebieten mit Erfolg durchführen liess: In einem von der Aussenwelt abgeschlossenen physikalischen System gibt es eine Summe von Energien, welche bei allen auftretenden Veränderungen konstant bleibt.

Nun zur Erhaltung der Masse. Die Masse ist definiert durch den Widerstand, den ein Körper seiner Beschleunigung entgegensetzt (träge Masse). Sie wird auch gemessen durch das Gewicht des Körpers. Dass diese zwei so verschiedenen Definitionen zu derselben Masszahl für die Masse eines Körpers führen, ist eine an sich höchst verwunderliche Tatsache, deren tiefere Bedeutung erst durch die allgemeine Relativitätstheorie aufgeklärt wurde. Der Satz lautet: Die Massen bleiben bei irgend welchen physikalischen (und chemischen) Veränderungen ungeändert. Die Masse schien so die eigentlich wesentliche Qualität der Materie zu sein. Bei Erwärmung, Schmelzen, Verdampfen, Auflösen, beim Eingehen chemischer Verbindungen ändert sich die Masse (bezw. die Gesamtmasse) nicht. Diesen Erhaltungssatz der Masse, dem die Physik bis vor einigen Jahrzehnten exakte Gültigkeit zugeschrieben hat, wurde durch die spezielle Relativitätstheorie als unzureichend erkannt. Er wurde durch diese Theorie mit dem Energiesatz verschmolzen, wie etwa 60 Jahre früher der Erhaltungssatz der mechanischen Energie und der der Erhaltung der Wärmemenge miteinander verschmolzen wurden.

Besser könnte man sagen: der Satz der Erhaltung der Energie hat ehedem den der Erhaltung der Wärme und neuerdings den der Erhaltung der Masse geschluckt und hat so allein das Feld behauptet.

Den Satz von der Äquivalenz von Masse und Energie pflegt man (etwas ungenau) durch die Formel auszudrücken

$$E = mc^2,$$

wobei c die Lichtgeschwindigkeit (3 · 10¹⁰ $\frac{cm}{sek}$) bedeutet. E ist die Energie, welche in einem (ruhenden) Körper steckt, in seine Masse. Die Energie, die zu der Masse m gehört, ist gleich dieser Masse, multipliziert mit dem Quadrat der ungeheuer grossen Lichtgeschwindigkeit, also ein gewaltiger Betrag pro Masseneinheit.

$\Lambda - 148$

ABOVE A young Albert Einstein.

With his Serbian wife, Mileva, who had studied physics with him at the ETH (a remarkable feat at that time), Einstein rented a flat on the second floor of 49 Kramgasse – now a museum – which became a forum for discussion of his ideas over eggs and sausages. His old friend Michele Besso, whom Einstein mocked as an "awful schlemiel", had a talent for asking childlike questions that were deceptively challenging and instructive. Years later, Einstein revealed how he had approached Besso to discuss his misgivings about his theory of an apparently unchanging speed of light, a foundation stone of relativity. It had been during one of their discussions, which often arose during the stroll between the Patent Office and Kramgasse, when he declared: "I could suddenly comprehend the matter."

Einstein would also hone ideas with his "students" in Berne, Conrad Habicht and Maurice Solovine. The three constituted themselves with mock formality as the Olympia Academy, with Mileva as an "observer". To Habicht Einstein described the first of his 1905 papers, which explained the photoelectric effect, a topic he had written rapturously to Mileva about, as "very revolutionary". On 17 March, three days after his twenty-sixth birthday, he sent it to the journal *Annalen der Physik*.

ANNUS MIRABILIS

Experiments had shown that electrons were ejected from metal surfaces when the surfaces were struck by light. But scientists of the day were puzzled as to why the speed with which the electrons emerged varied with the colour rather than the intensity of the light. Einstein suggested that light consisted of particles, which he called light quanta. As the brightness increased, more quanta rained down on the metal and more electrons were blasted out. But the speed with which they emerged only increased when the quanta grew bigger, when they moved from red to the blue end of the spectrum.

Einstein was returning in part to Newton's idea of light as a stream of particles, long since dropped because waves are needed to explain such effects as interference and diffraction. Yet even as Einstein talked of light quanta, or photons, as they were later named, he also referred to light as having a frequency, an essential part of wave theory. He was confronting what has become a famous paradox: that light has the properties of both waves and particles. In 1922 Einstein's explanation of the photoelectric effect won him a belated Nobel Prize.

Einstein's next significant paper in 1905 was received by *Annalen der Physik* on 11 May. Its subject was Brownian motion, named after Robert Brown, the Scottish naturalist who first used a microscope to observe the random motion in water made by tiny particles such as fragments of pollen grain. What caused this zigzag movement had puzzled Brown, but Einstein knew the answer: the particles were being buffeted by invisible molecules. At a time when some physicists still doubted the physical reality of atoms this was an important insight.

WRESTLING WITH SPECIAL RELATIVITY

Perhaps Einstein's greatest paper that year, on "special relativity", was received by the journal on 30 June 1905. Its aim was to reconcile Newton's laws of motion with the theory of electromagnetism developed in the 1870s by one of Einstein's heroes, the Scottish mathematician James Clerk Maxwell. Einstein had wrestled with the ideas that would lead to special relativity since the age of 16, when he wondered what it would be like to ride a beam of light, where light is electromagnetic radiation, an electric field that oscillates in both time and space.

Travel at light speed was allowed by Newton's laws, but this did not square with Maxwell's description of light. His equations describe the classical behaviour of the electromagnetic field and predict the speed, c, with which electromagnetic waves such as light propagate. But if we regard a light beam as a series of troughs and peaks, then we can see that an observer moving alongside with the same speed would keep pace with a particular trough or peak, and would no longer "see" the oscillation.

There was a deeper problem. According to Newton's laws, there is no such thing as absolute motion. If you measure the speed of a train, it could be relative to the ground, relative to the Moon or relative to some distant galaxy. At first sight this principle is inconsistent with Maxwell's prescription of an absolute value for the speed of light.

What Einstein did was to accept these contradictions. To keep the speed of light constant – irrespective of the speed of the observer relative to the source of the light – Einstein had to distort time and distance, predicting head-spinning relativistic effects, for instance that a moving clock would seem to tick more slowly than one at rest.

In the September of his annus mirabilis, Einstein submitted a brief article in the *Annalen der Physik* on the relationship between mass and energy, which showed that when a body releases energy in the form of radiation its mass decreases by a proportionate amount. But the true importance of this finding did not become clear until two years later, when Einstein announced that the reverse was also true: that all mass has energy, as expressed by his equation $E=mc^2$.

$E = MC^2$

Einstein's theory predicted that a vast amount of energy can be extracted from a tiny amount of mass – such as the loss in mass that occurs when the nuclei of heavy elements like plutonium fall apart, in the case of a nuclear weapon. Einstein called this the most important consequence of his relativity theory, and the atom bomb was to provide the most awful confirmation of the theory – although he had nothing to do with its development.

When we consider the extraordinary body of work that Einstein published in 1905, many are struck at how a serious journal like *Annalen der Physik* took a minor official from the Patent Office seriously, something which would not happen today. Einstein was certainly fortunate in that he'd already got his first (and not very impressive) papers published (one on capillarity in early 1901 and the other on dilute salt solutions in 1902), and the journal had a policy of publishing subsequent work.

At first glance, the 1905 papers seem to cover disparate topics, but they are united by Einstein's "particulate" view of the world, understanding it in terms of atoms, molecules, electrons and tiny chunks of energy – "quanta" – which was revolutionary. In his relativity paper, for example, where he meshed theories of movement and electromagnetism, he viewed electricity as electrons.

EINSTEIN'S LARGE, SHINING EYES

Though Einstein's creative explosion in 1905 marked the start of a remarkable two-decade run at the cutting edge of physics, when many picture him today, they conjure up images from his latter years in Princeton, of a little professor with electrified grey hair shuffling around without socks, puffing his pipe. These are the images of Einstein that adorn Hawking's office. But the 26-year-old Einstein was sociable and funny, a good-looking and athletic Bohemian who had "extraordinarily large, shining eyes", according to one of his female admirers.

Why, then, did the elderly Einstein leave such an indelible mark on the public consciousness? One key reason was the media brouhaha that followed the confirmation in 1919 by a British team led by Sir Arthur Eddington of his greatest theory, general relativity, which overturned Newton's picture of gravity. Eddington and his team measured the bending of starlight during a solar eclipse, thus confirming the curvature of space-time as a result of the influence of the motion of massive bodies, just as Einstein had predicted.

After the First World War's senseless, nationalistic slaughter, the idea of a German theory being backed by a British expedition led by a Cambridge professor marked an inspirational vindication of rationality. "Lights all askew in the heavens, men of science more or less agog … Einstein theory triumphs", announced the *New York Times*.

The Einstein legend was born, and the resulting public fascination with him endured for decades. If anything, Einstein's influence on physics is stronger today than ever before, creating sturdy scientific foundations on which many other scientists have built their research. As became the case for Stephen Hawking himself, the more Albert Einstein advanced in his career, the more he was interested in finding a unified theory. Unfortunately, he failed to achieve that grandiose objective, as have all those who have followed in his giant footsteps.

SEE ALSO:

A Brief History of Time, p.104
Stephen Hawking, You Are Wrong!, p.176

Einstein was confronting what has become a famous paradox: that light has the properties of both waves and particles.

Stamp of Authority

These stamps were used by Stephen Hawking's assistants to authenticate documents. While some also needed his thumbprint, the one in the middle bears a copy of his final signature.

"That was actually the last actual signature, preserved for all time, that Stephen actually did himself," commented his former PA, Judith Croasdell. "He would use it when writing to fans, such as Laya Sarakalo." The stamp was also used to sign documents, for instance when he gave his first synthesizer on loan to the Science Museum in 1999.

Hawking had lost the use of his hands for writing in the early 1970s, with the final, complete loss occurring during his 1974–75 year, according to his collaborator and friend, the Nobelist Kip Thorne. After he could no longer write equations, Hawking developed a remarkable skill to conjure up geometrical and topological images of mathematics in his head. Moreover, as I discovered when he signed my own copy of *Universe in a Nutshell*, Stephen Hawking could "sign" a book or a document with the help of an ink pad, his thumb and his then wife, Elaine.

His thumbprint was also made into a stamp. To make his signature official this stamp was used: "Right thumbprint of S. W. Hawking witnessed by..." Other stamps bore the name of a personal assistant, such as Judith Croasdell, who can remember using hers for tax and legal affairs, and to sign books: "For ten years as PA it was my signature on hundreds of books with the official stamp ... only the PA had the official stamp decreed by the University."

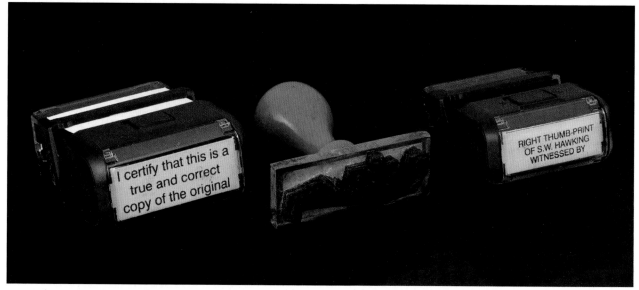

ABOVE Among several stamps in the collection is one that bears Stephen Hawking's last written signature (shown centre).

After he could no longer write equations, Hawking developed a remarkable skill to conjure up geometrical and topological images of mathematics in his head.

Picturing Hawking

It is unusual for a scientist to sit for so many portraits, and perhaps even more so when that person is disabled in some way. Beginning with remarkable David Hockney drawings that date from the 1970s, the numerous depictions of Stephen Hawking add up to an unparalleled visual record that tracks the effects of his fame, ageing and motor neurone disease.

HOCKNEY DRAWING HAWKING DRAWING HOCKNEY

In 2011, during the early discussions about the Science Museum's exhibition for Hawking's seventieth birthday, I was surprised to learn that David Hockney had drawn portraits of the scientist as long ago as 1978. As the museum's Hawking curator, Juan-Andres Leon, commented, "It is remarkable that Hockney made a portrait of Hawking long before he was a celebrity." It turned out that, as a rising star of pop art, Hockney had been commissioned by Stephen Hawking's college, Gonville & Caius.

Hockney produced sparse line drawings likened by some to Picasso's 1920s depictions of individuals such as his wife, the ballet dancer Olga Kokhlova, and the composer Igor Stravinsky. One shows him with a thick tie, checked shirt, no glasses, and hands crossed. The latter feature was particularly poignant, as a few years earlier, in 1974–75, he had lost the ability to write equations. Another portrait also showed his daughter Lucy reading what Hockney told me in an email was a *Star Wars* book. "I remember it very clearly with his blond hair and dark glasses, sitting on the sofa drawing my dad," she said. "He has taken artistic licence when it comes to the book, however," Lucy added. "At that time, when I was seven, I drew a picture of David Hockney as he was drawing my dad. I proudly signed my picture, gave it to him and told him, 'I'm an artist too.'"

Over the following decades, Hawking's growing fame and achievements saw him become the subject of numerous other portraits by distinguished artists. As with any subject, the challenge for each was how to capture the spirit of this remarkable scientist in two dimensions, with the added consideration of portraying his disability in an authentic way, yet with sensitivity.

In the 1980s the National Portrait Gallery, encouraged by Margaret Gowing, a historian of science and a trustee of the gallery, chose Hawking as a subject as part of a project to raise the profiles of contemporary British scientists. Painted by Yolanda Sonnabend in 1985, Hawking's face looks strangely feminine as he sits upright in front of a blackboard, without glasses and with his hands crossed. Lucy Hawking said that her father disliked the portrait so much that he asked the National Portrait Gallery to remove it from public display: "She insisted on doing the portrait without his glasses on, when he always wore them, and he felt that she homed in on his disability rather than him as a person."

DONS AT DESSERT

Within the collection of oil paintings owned by Gonville & Caius College in Cambridge there is a striking 1989 portrait of Hawking by the artist Paul Gopal-Chowdhury. He had depicted Hawking before, as part of a group of starkly lit figures in the 1984 work Dons at Dessert, which also hangs in Gonville & Caius. Hawking is sitting near the back of the group and easy to miss. That year Gopal-Chowdhury had been the college's artist in residence, and the dons had all dropped in to pose for him in pairs.

The challenge for each artist was how to capture the spirit of this remarkable scientist in two dimensions, with the added consideration of portraying his disability in an authentic way, yet with sensitivity.

For his later portrait of Hawking, Gopal-Chowdhury recalls how Stephen's assistant had to hold his head up – "he had no muscles in his neck." Since Hawking could not hold a pose, Gopal-Chowdhury also worked from a photograph and drew Stephen in his wheelchair.

HAWKING'S DEVIL-MAY-CARE SPIRIT

Bespectacled, with a direct gaze at the viewer, hand poised on the controller of his speech synthesizer, the Hawking in this 1989 portrait looks as if he is about to begin a seminar. "Stephen was very pleased with it," said Gopal-Chowdhury, adding how he was inspired in part by seeing Hawking drive his wheelchair in traffic along Madingley Road in Cambridge. He wanted to capture Hawking's defiant, devil-may-care spirit, rather than portray someone who seemed "defeated and down".

When Hawking sat for Frederick Cuming in 2005, he himself had given the commission, recalled Cuming, who died in June 2022. By now Hawking was the best-known scientist in the world. "He was always busy teaching students," Cuming remembers, "so was not available for a proper sitting. My wife and I visited Stephen on several occasions at his home. My studies consisted of a number of drawings made in situ, and from them I produced a number of paintings. I always work on a mid-tone background so that dark and light areas register the structure and composition of the painting."

Painted in fragmented brown brushstrokes, the resulting portrait shows Hawking deep in thought, gazing off to his right, his head now resting as his neck muscles found gravity increasingly hard to defy. "In contrast to the Sonnabend portrait, which depicts a young man at the outset of his reputation", comments the National Portrait Gallery, "Cuming's portrait, with its tenebrous quality, evokes the sitter in his maturity."

"The Cuming portrait is the one he liked most," said Juan-Andres Leon, "and he was friends with the artist up until his death." The Cuming, added Hawking's daughter Lucy, "was the portrait we wanted to be shown at the National Portrait Gallery".

Hawking's cosmic, inquiring gaze is restored in Tai-Shan Schierenberg's tondo (circular) portraits of Hawking. One hung in the Department of Applied Mathematics and Theoretical Physics in Cambridge, and the other in the Royal Society. They were both painted in 2008, a couple of years after Hawking began using muscles in his cheek to control his voice synthesizer, and were commissioned by the computer pioneer and philanthropist Dame Stephanie Shirley.

For the planned exhibition for Hawking's seventieth birthday in 2012, it struck me and curator Boris Jardine that it would be interesting to ask David Hockney to return to his early subject. Given the artist's long flirtation with technology, which has seen him work with multi-screens, high-definition video, colour photocopiers, fax machines and an iPad, it seemed natural to ask him to use a different medium for his new work. We were already planning to show that rarely seen Hockney portrait, dating from 1978, owned by Hawking's first wife, Jane. What about an iPad portrait too?

Both Hockney and Hawking were excited by our idea. Arrangements were made for them to meet before the opening of Hockney's triumphant A Bigger Picture exhibition at the Royal Academy, but they had to be put on hold as Hawking fell ill, missing also his birthday celebrations in Cambridge and at the museum. The family were worried that it was morbid to do a portrait, given that Hawking was so ill, but, after a great deal of to-ing and fro-ing, the sitting eventually went ahead.

For his iPad art Hockney used an app called Brushes, which allows the artist to work faster than with watercolour, or even coloured pencils. The software allows for the use of thumbs and fingers or a stylus, and offers the ability to modify the hue and colour and layer brushstrokes of various widths and opacities.

On a screen, we were able to show visitors to the exhibition how the portrait was created in real time, with what was in effect a recording of how Hockney captured Hawking stroke by stroke. The work begins at the top of Hawking's head on a beige background. A basic sketch of Hawking's bespectacled face peers out early on, adorned with violet eyes. Pencil-like strokes add detail, and paint-can sprays fill in his cobalt suit, a light blue cravat, computer screen and shadows.

A POLKA-DOT CRAVAT

After a while, Hawking's face gets a more lifelike hue, polka dots appear on his cravat and the broader contours of his wheelchair emerge. His hands are moved, and a joystick, green background and overhead light are added before Hockney returns to work on his face. The artist repeatedly plays with shading and skin tone before the final portrait of the world-famous physicist appears in its finished form. "He really liked it," said Lucy.

Seeing the iPad portrait emerge next to the 1978 line drawing offered an intriguing comparison: the technology is so different but, whether paper or a digital pad, it's Hockney's draughtsmanship and Hawking's instantly recognizable face that are the highlights. It was a thrill in 2012 when Hockney rang me to join him at the museum so he could see his iPad portrait created, digital stroke by stroke, in our Hawking exhibit.

Stephen Hawking was depicted in ways other than pencil, paint and pixels. His office also contained a resin

ABOVE Hawking commissioned a series of portraits by the British painter, Fred Cuming. This painting was kept at Hawking's home; another example was given by the scientist to the National Portrait Gallery in 2006.

"The Cuming portrait is the one he liked most ... and he was friends with the artist up until his death."

The artist repeatedly plays with shading and skin tone before the final portrait of the world-famous physicist appears in its finished form.

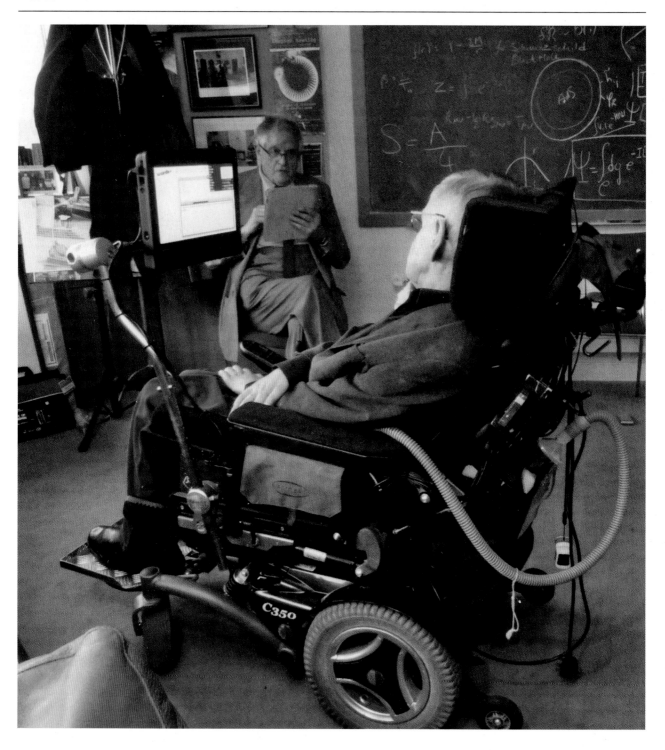

OPPOSITE David Hockney sketching Hawking, taken by his assistant at the time, Judith Croasdell. **ABOVE** Hockney's 2012 iPad portrait of Hawking.

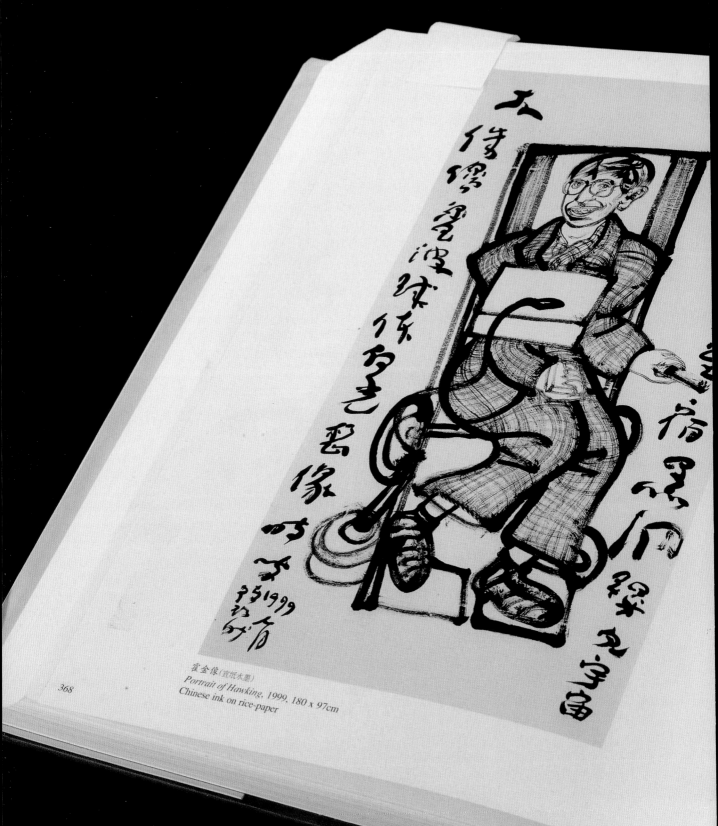

霍金像 (宣纸水墨)
Portrait of Hawking, 1999, 180 x 97cm
Chinese ink on rice-paper

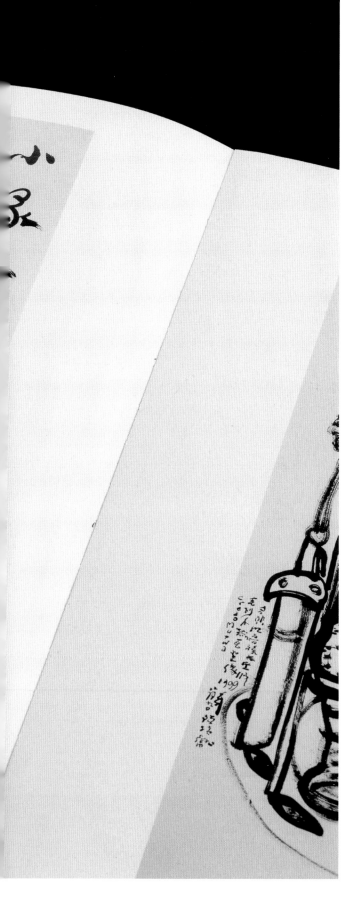

portrait by Ellie Shuster, dated 2016, who had sculpted him in clay and reproduced the bust in cold-cast bronze.

Another of his more unusual portraits can be seen on a medal for science communication that was unveiled in the Royal Society in 2015. The sketch on which it was based was created by the cosmonaut Alexei Leonov, who had presented his drawing to Hawking in the Science Museum's IMAX theatre in 2015.

Leonov was remarkable because he had prepared for a Soviet moon landing to rival Apollo in a 5-metre LK-3 lunar lander, which was never used and went on temporary display in the Science Museum's Cosmonauts exhibition that they had both helped to launch in 2015. Most famously, Leonov was the first spacefarer to perform a spacewalk, a feat which almost ended in tragedy: because of the vacuum of space, his spacesuit had not only expanded and stiffened, but was also now too large for re-entry through the airlock into his spacecraft (as a desperate last resort, Leonov opened a valve to deflate his spacesuit). The reverse side of the Hawking medal shows Leonov during this feat, when, as he told us at one Science Museum event, he was struck by the sound of his own breathing, his heartbeat and a sense of the universe "being limitless in time and space".

Given Hawking's fascination with space travel, it was fitting that he was depicted by Leonov: the cosmonaut had also become the first person to create art in outer space, with a small yet remarkable view of a sunrise.

SEE ALSO:

Royal Recognition, p.102

OPPOSITE Book with hand drawings by the Singaporean artist Tan Swie Hian, including Stephen Hawking. Nanyang Academy of Fine Arts, 2001.

Blackboard of Scientific Doodles

As you stood in Stephen Hawking's office, it was hard to miss this blackboard, adorned as it was with strange doodles, cartoons and equations.

To the uninitiated visitor, like me, the graffiti, puns and in-jokes were mostly meaningless. But this enigmatic blackboard had a special place in Stephen Hawking's heart. Initially removed as a memento of a momentous gathering by his then personal assistant, Judy Fella, he had preserved it for almost four decades, even bringing it with him when he moved from his old office in Silver Street to have it mounted on the right-hand side of his desk.

The chalk scribbles hold a record of a meeting Hawking had organized in 1980, the Nuffield Workshop in Cambridge on Superspace and Supergravity. His colleague, Martin Rees (now Lord Rees, the Astronomer Royal), had secured the funding from the Nuffield Foundation and, according to Hawking's biographer Graham Farmelo, Hawking "jumped at the opportunity to explore supergravity, which was the 'hottest new theory in fundamental physics'."

SUPERSYMMETRY

"At that time, people who were interested in fundamental physics worked on either gravity or particle physics, the forces that held together atoms," says Farmelo. "Hawking was very much a gravity person but, to his credit, he grabbed at a theory that had emerged from the world of particle physics and tried to incorporate supersymmetry into our understanding of gravity."

Supersymmetry, a concept nicknamed SUSY and developed by particle physicists in the latter half of the twentieth century, suggests that the two basic types of particles that make up our world – fermions, the

RIGHT Blackboard inscribed with doodles and drawings made during the 1980 Nuffield Workshop in Cambridge on Superspace and Supergravity. Hawking kept this blackboard as a treasured souvenir of the international conference at which he and his colleagues thought they were on the verge of a "theory of everything".

The theory attempted nothing less than to unify all the four fundamental forces of nature, and marked a significant extension of Einstein's theory of gravity, otherwise known as general relativity.

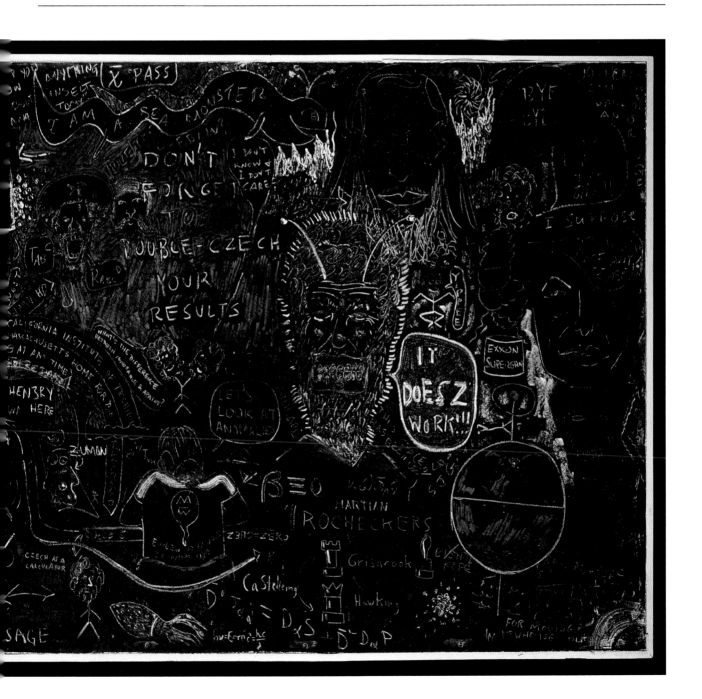

particles of matter such as electrons and quarks, and bosons, the force-carrying particles such as photons and gluons – are merely two aspects of a single particle. By this view, early after the Big Bang that created our universe, a single type of particle split into these two families of particles in a process of "symmetry breaking", like the branching of a great river of creation.

Every known particle therefore, has, a supersymmetric partner, that is one with symmetry with respect to transformations in time and space (see page 84). As Farmelo explains in *The Universe Speaks in Numbers*, this is the "last great unfulfilled symmetry". Physicists at the time believed that the theory of supersymmetry would allow them to explore beyond what was known as the Standard Model of particle physics: the tried-and-tested model that accounted for three of the four known fundamental forces – electromagnetic, weak and strong interactions – but omitted gravity.

SUPERGRAVITY

Enter supergravity, which the Nuffield conference of 1980 set out to discuss. Key contributions were made at the State University of New York, Stony Brook, by Peter van Nieuwenhuizen and Daniel Freedman, which suggests that the particle that causes gravity (a graviton) ought to have a partner (called a gravitino). The theory attempts nothing less than to unify all the four fundamental forces of nature, and marks a significant extension of Einstein's theory of gravity, otherwise known as general relativity.

The full list of participants at the Cambridge conference reads like a *Who's Who* of the entire generation of physicists working on supersymmetry. Both van Nieuwenhuizen and Freedman attended, along with Sergio Ferrara, who had worked with Freedman at the École Normale Supérieure in Paris, and is seen as the third main architect of supergravity – in 2019, the three were awarded the highly regarded Breakthrough Prize in Fundamental Physics, which, unlike the Nobel Prize, has a record of rewarding ideas even if they lack experimental verification.

A GIDDY TIME FOR PHYSICS

At the conference, Hawking and his colleagues came to think that with supergravity they had achieved a quantum theory of gravity – a first draft of a theory of everything, extending the work Stephen Hawking had begun with his stunning 1974 paper on why black holes "ain't so black". According to Farmelo, he was trying to work with two apparently incompatible theories: the quantum field theory of fundamental particles and Einstein's general relativity. He was a world authority on the latter but knew little about the former. Even so, "somehow he found a way through – amazing!"

They now seemed on the verge of combining two cornerstones of theory: general relativity, which describes the universe on the largest scale, and quantum theory, which explains the very small. So, this blackboard represents a giddy time for physics, and it's no surprise that when the conference proceedings were published as a book, a photo of the blackboard even adorned the cover.

During the 1980s some of the ideas on supergravity discussed at the Cambridge conference came to be used to develop superstring theory, a version of string theory – in which particles are represented as vibrating one-dimensional objects called "strings" – that incorporates supersymmetry.

Both supergravity and superstrings can be seen as an extension of another quest dating back more than a century, when mathematical symmetry provided a powerful organizing principle for the laws of nature, the rules – usually expressed as mathematical equations – that try to explain everything we observe in the universe. Initially, explains one of Stephen Hawking's students, Thomas Hertog, supergravity and superstrings were seen as "competing, even orthogonal" (symbolically at right angles to each other) theories, and it was only later, in the 1990s, that they were shown to be related. Indeed, they mark a nice example of how theories can converge, and even be unified: supergravity turns out to be equivalent to string theory when describing low-energy interactions. In a similar way, Newton's theory of gravity gives the same

Hawking preserved the blackboard for almost four decades, even bringing it with him when he moved from his old office in Silver Street to have it mounted on the right-hand side of his desk.

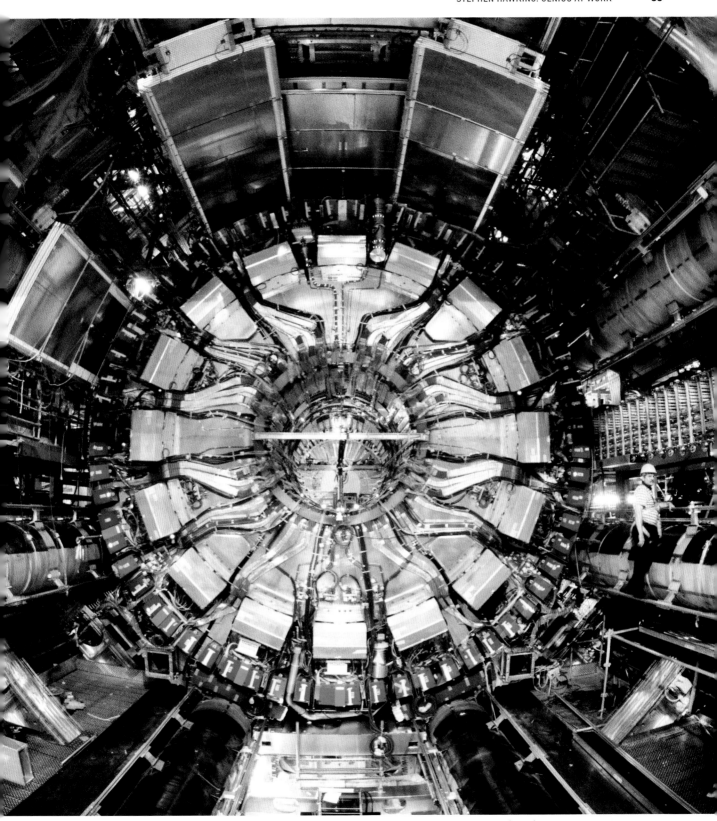

ABOVE One of the vast detectors of the Large Hadron Collider, at the European Organization for Nuclear Research (CERN) in Geneva.

answers as Einstein's in the relatively low-gravity environs of our Solar System.

THE SEARCH FOR SUPERSYMMETRY

Until the 2010s, researchers working on theories based on supersymmetry, including superstrings and supergravity, were hopeful that experiments in a new generation of particle accelerators would confirm them, notably the 17-mile-circumference Large Hadron Collider (LHC) at CERN, the European particle-physics centre near Geneva. Sure enough, at the LHC in 2012 the Higgs boson was discovered, meaning that all known particles are encompassed within the Standard Model of particle physics, which was now complete. This much heralded discovery validated theoretical work that Peter Higgs, along with the Belgian François Englert and the American Robert Brout, had done with pen and paper half a century earlier. But it came as a disappointment to Stephen Hawking, who had bet $100 that the Higgs boson would not be found at the LHC, to make the point that its absence would actually make for more interesting physics.

After more than a decade of operation, however, the giant LHC has not found any evidence of supersymmetry. At the time of writing, previously popular supersymmetric theories that attempted to solve problems like dark matter, a mysterious source of gravity, and the fine-tuning of the Higgs boson seem to have not worked out as they had predicted particles that the LHC should have seen by now, according to Harry Cliff, a particle physicist who often encountered Hawking at university and has worked on Science Museum exhibitions.

But he added that it is a "glorious failure" in the sense that, though it has not worked to date, the theory is silent about the energies at which supersymmetric strings would become apparent, and it could well be that much higher energies will be needed than are currently accessible. In turn, however, the very fact that superstring theories are unable to furnish predictions amenable to test by the current generation of particle colliders has sparked heated debate about whether string theory counts as "real science" at all.

Symmetry in Science

Symmetry is a central idea in twentieth-century physics. We are familiar with the idea of symmetry in everyday life: the exterior of our bodies is essentially symmetrical. But there are many other symmetrical shapes in nature, from snowflakes to sunflowers.

However, mathematicians formulate this deceptively simple idea in a precise way: they say an object is symmetric if it looks the same after what is called a transformation. A butterfly's left wing is the mirror image of its right: therefore, it has reflectional symmetry. The right side of a human liver is not the mirror image of the left: hence it does not have reflectional symmetry. A flower can be rotated and yet the petals look the same afterwards: it has rotational symmetry. A sphere has both kinds of symmetry, reflectional and rotational.

The same thinking can be applied to time, too: many theories are indifferent to the direction of time, and are time symmetric. The Second Law of thermodynamics is not. Moreover, there is "time violation" (meaning violation of time reversal –

reflection – symmetry) in the decay of some particles. Mathematicians also apply their definition of symmetry in a more abstract way, as immunity to the impacts of change. For example, if you like grapefruit juice just as much as you like orange juice, then your preference is "symmetric" under the transformation that exchanges grapefruit and oranges.

We now know that symmetries – the changes that do not change anything – pervade the laws of physics, which is why, for example, equations don't change in different places or different times. In 1915 this principle led to a profound link being established between symmetry and the laws of nature, by the German pure mathematician Emmy Noether, a pioneering (and at that time rare) woman in the field.

Noether set out to resolve a puzzle in Albert Einstein's newly minted theory of gravity, the general theory of relativity. According to the theory, it seemed that energy might not be conserved. To fix the problem, Noether showed that the symmetries of general relativity ensure that energy *is* always conserved: a seminal idea. To do this, she teased out a link between two important concepts in physics:

DON'T FORGET TO DOUBLE-CZECH YOUR RESULTS!

Many of the jokes on Stephen Hawking's blackboard refer to the co-organizer of the Nuffield conference, Martin Roček, who was born in Czechoslovakia, where his father was a renowned chemist at the Czechoslovak Academy of Sciences. In 1960 his family made their escape from the Eastern bloc to the United States, and in due course Martin became a theoretical physicist, doing his degrees at Harvard University. In 1979, to find out more about supergravity, Hawking had hired Roček, then studying for his doctorate.

"At that time," recalled Roček, "he was already confined to a motorized wheelchair and could not write himself." Blackboards, therefore, had become central to how Stephen worked with his doctoral and postdoctoral students. "He could still speak," Roček went on, "albeit with such difficulty that only those who spent a lot of time with him could understand him. He would work with them by asking them to write equations on the blackboard."

IT DOESZ WORK!!!

In a similar way, said Roček, every time the delegates at the Nuffield conference hit an impasse in their discussions, they would doodle on the blackboard. Roček is depicted as a shaggy-bearded Martian with antennae, bellowing, "IT DOESZ WORK!!!"

Some of the scribbles from the conference are simple words that are easy to decipher: "Czech" refers to Martin, of course, "ZUMON" to the Italian-born theoretical physicist Bruno Zumino, who in the early 1970s had helped develop supersymmetry at the European Organization for Nuclear Research (CERN) in Geneva, and with the American physicist Stanley Deser constructed supergravity, separately from Freedman, Ferrara and Van Nieuwenhuizen. The latter are namechecked on the blackboard as "Peter and Jan", together with another Dutch physicist, Jan Van Holten.

Other doodles are of mysterious creatures, representing the group of mathematical functions or operators known as "Vielbein" – *vielbein* being German for "many-legged". Hence, for example, an Einbein has one leg, a Vierbein has

symmetries and conservation laws, whereby a conservation law – the conservation of energy, for example – states that a particular quantity must remain constant, such that energy can't be created or destroyed, for example.

EINSTEIN'S MUSE

Noether's powerful theorem says that for every conservation law, there's an associated symmetry, and vice versa – and that every such symmetry has an associated conservation law. She had revealed the mathematical reasoning behind these laws: energy conservation comes from translation symmetry in time, momentum conservation from translation symmetry in space, and the conservation of angular momentum from rotational symmetry.

When she died in 1935, Einstein himself wrote a letter to the *New York Times* in which he called her "the most significant creative mathematical genius thus far produced since the higher education of women began". Noether and symmetry have both occupied centre stage in physics ever since, though she has not enjoyed the wider recognition she so richly deserves.

ABOVE German pure mathematician Emmy Noether, who set out to resolve a puzzle in Einstein's general theory of relativity in 1915.

ABOVE Believed to be a depiction of Hawking himself, shown from behind and saying "Let's look at anomalies."

four, and Achtbein, the most promising for these "stupor symmetry" theories, has eight and appears under different guises. Why, therefore, do we find on the blackboard what appears to be a floppy-nosed squid climbing over a brick wall? The answer is that this "squid" is an elaboration of the "Kilroy was Here" meme that dates from the Second World War, in which a pudgy face is shown peering over a wall. Here Kilroy is replaced with an Achtbein.

THE DAMTP SQUID
The conference delegates' doodles could be as baffling as the mathematics they used. The reference to "Exxon supergrav" remains obscure. Other references are more straightforward, such as the speech bubble with "It doesn't make any sense!" or the sign that reads "Townsend" and "West", the names of two of the participants. The "damtp squid" is clearly a nod towards Hawking's own Department of Applied Mathematics and Theoretical Physics.

The only person of colour present was physicist Jim Gates of the Massachusetts Institute of Technology: "Someone decided to draw me as a man with what looks like a fishbowl on my head," he remarked in a documentary (*Hawking: Can You Hear Me?*). "It was actually my afro."

Where is Stephen Hawking himself on the blackboard? "I think it is rather plausible that Stephen is underneath the balloon," suggests his Cambridge colleague and fellow attendee, Gary Gibbons, "saying, 'Let's look at anomalies.' His glasses and hair are convincing." The "anomalies" in question, he goes on, were to do with broken symmetries in quantum field theory, the theory that sets out to blend quantum mechanics and elements of the theory of relativity.

Another subtle reference to Stephen Hawking can be found in the chess figures on the blackboard, adds Gibbons. A king and rook are castling, a reference to

ABOVE Sergio Ferrara, Peter van Nieuwenhuizen and Daniel Freedman at the 2020 Breakthrough Prize at the ASA Ames Research Center, Mountain View, California.

Blackboards, therefore, had become central to how Stephen worked with his doctoral and postdoctoral students.

three of the workshop attendees: Hawking (acknowledged as "the king" of cosmology), Mark Grisaru ("GrisaRook") and Kelly Stelle ("CaStelleing"). "Bishop" is crossed out and replaced by "Pope": Chris Pope, a research student of Stephen's.

Today, the overarching message of this blackboard is that physics back in 1980 was not as advanced as Stephen Hawking had thought during that heady meeting in Cambridge. So why did he treasure it for so long? According to Martin Roček, "He had a great sense of humour."

SEE ALSO:

Royal Recognition, p.102

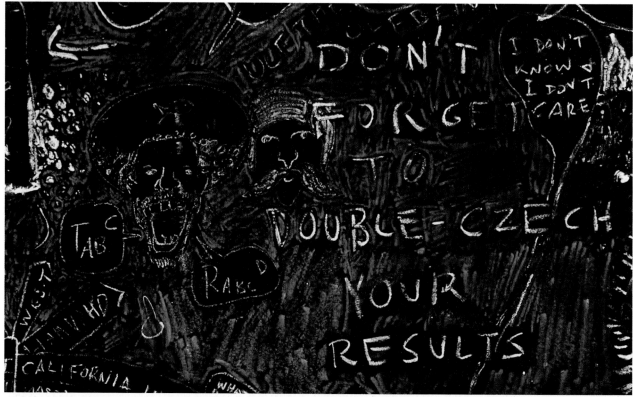

ABOVE "Czech" is a reference to the co-organizer of the Nuffield conference Martin Roček, who was born in Czechoslovakia.

ABOVE A floppy-nosed squid climbing over a brick wall, an elaboration of the "Kilroy was Here" meme that dates from the Second World War, in which a pudgy face is shown peering over a wall. Here Kilroy is replaced with an Achtbein.

ABOVE Martin Roček depicted as a shaggy-bearded Martian with antennae, bellowing, "IT DOESZ WORK!!!"

Hawking's Voice

The story of Stephen Hawking's voice is long, complex and full of surprises, not least how he managed to become the greatest science communicator of his age after his natural voice was permanently silenced.

In 1983, when he appeared on the BBC's *Horizon* science programme, his speech was already slurred and hard to decipher. Just two years later, aged 44, his voice disappeared altogether in the wake of a visit to a science conference in CERN, the European Organization for Nuclear Research, in Geneva.

During the trip, Stephen had contracted pneumonia. He was put on a ventilator, and became so seriously ill that his wife Jane was asked for permission to turn off his life support. A Christian, she vehemently refused, and he was flown to Addenbrooke's Hospital in Cambridge, where the doctors managed to curb the infection. But his motor neurone disease was beginning to paralyse the muscles of his voice box – his larynx – blocking his airway, so, to help him breathe, they performed a tracheotomy, which involved cutting a hole in his neck and placing a tube into his windpipe. This left him permanently unable to speak: now, the only way he could communicate was by raising his eyebrows to select letters as they were held up on cards. According to Jane, the impact was devastating.

But by the time I first encountered Hawking, in California in 1988, when he was promoting *A Brief History of Time*, he was using a voice synthesizer, which allowed him to speak again, in that paradoxical blend of machine and human, which had an impact far beyond communicating his science. This, according to his youngest son Tim, also enabled the start of a father–son relationship. However, his oldest son Robert described in a eulogy to his father how it had the opposite effect on their relationship: "When I was a teenager, he still used his own voice, which was very hard for most people to understand. I strongly valued that I could understand him very well. After his tracheotomy, he acquired his iconic word picking system and computer voice … I felt a sense of loss from the slower communication."

THE EQUALIZER

Hawking was using a computer program called Equalizer developed by a California company, Words+, which allowed him to select words and commands on a computer using a hand clicker on a slim black control box. From the first characters Equalizer could guess the most likely words, enabling Stephen to type in fewer characters, and offered shortcuts, so a special key could be used for sentences he would use every day.

Words+ CEO Walt Woltosz had come up with an early version of the program in 1980 for his mother-in-law, who also suffered from motor neurone disease and had lost her ability to speak and write. For Hawking, he recalled, "I made a simple version of Equalizer on an Apple IIc computer, because, at the time, Stephen was in an intensive care unit and it was the smallest computer with enough power to do the job." But customs held up the import of the little computer ("They couldn't believe it was free"), and by that time Stephen was out of ICU he could use a bigger computer, an IBM, capable of producing a voice output.

The Words+ setup allowed Hawking to write at 10–15 words per minute, and a voice synthesizer, initially a Votrax Type'n Talk, converted the input text to phonemes, then joined these up to create spoken words. "Phonemes are individual sounds, like *sss* for s, *eee* for e, *oh* for long o, *ah* for o, as in the word 'stop', *tuh* (without the uh sound) for t, *buh* (without the uh sound) for b, and so on," explained Woltosz. "So to speak the word 'stop' would require *sss, t, ah, p*." Later, synthesizers based on diphones, a unit of speech made up of two simple speech sounds, were developed that could combine a series of diphones, so stop might be "*st, ahp*", for example.

Stephen then moved to using a Speech Plus/ Telesensory Systems voice synthesizer board, the CallText 5000, thanks to the generosity of Telesensory Systems. Woltosz's UK distributor, David Mason, the husband of one of Stephen's nurses, Elaine (who would become Stephen Hawking's second wife), adapted this desktop system to become portable, as a grey box fixed to the back of his wheelchair.

Now Stephen Hawking's voice could accompany him wherever he went. He first used this synthesizer to deliver a talk at the Texas Physics Conference in Chicago,

OPPOSITE The first case built to contain Stephen Hawking's voice synthesizer. It would hang from the back of his wheelchair.

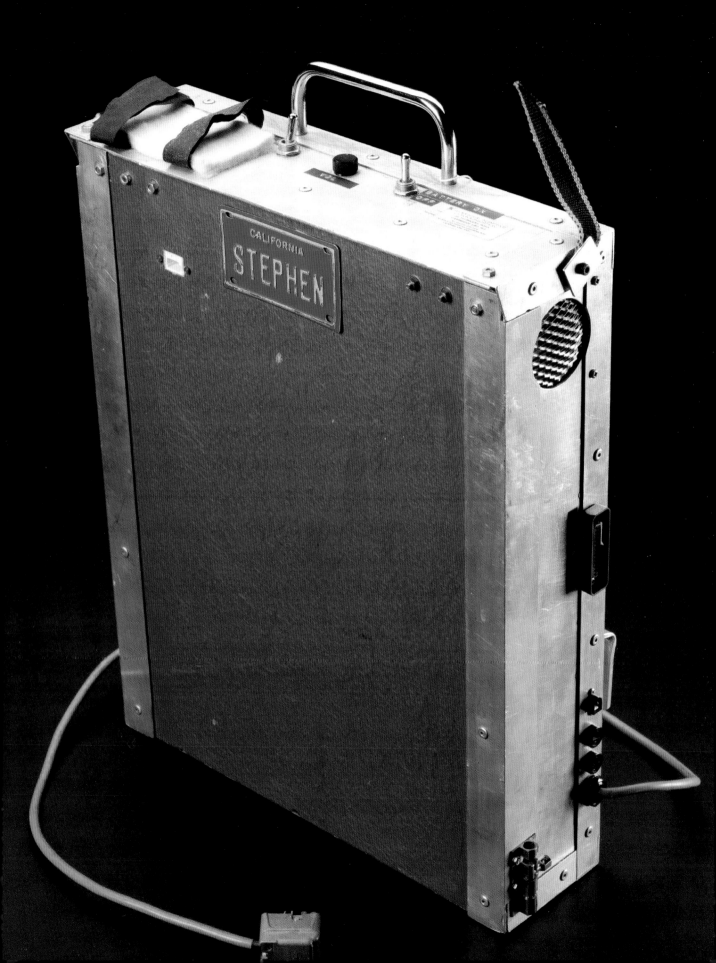

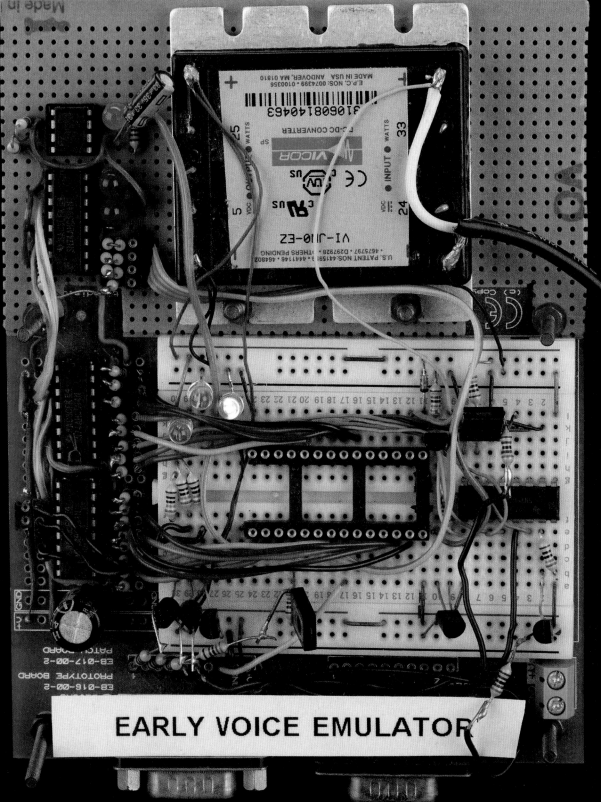

EARLY VOICE EMULATOR

chaired by John Wheeler, the American theoretical physicist who coined the term black hole. "It is the best I have heard," said Hawking, though it did have a side effect: "it gives me an accent that has been described variously as Scandinavian, American or Scottish."

"THAT'S NOT MY VOICE"
In 1988, Stephen visited Speech Plus in California while publicizing *A Brief History of Time*, and it looked as if his voice would be changed once again. But he was having none of it. In 1988, Walt Woltosz told me why Hawking had rejected the upgraded voice: "He didn't like it. He said, 'That is not my voice,' and made them put the old one back."

That remained the case until 2014, when the CallText 5010 hardware (a later generation of the CallText 5000) began to show wear and tear. Hawking's assistant Jonathan Wood feared that, if it failed entirely, his distinctive voice would be lost for good. He approached one of the original Speech Plus team, Eric Dorsey, in Palo Alto. "Dorsey tracked down two additional voice synthesizer cards found in the attics of his then colleagues at Speech Plus," recalled Wood, "and arranged for them to be donated to Stephen. He also tracked down the firmware to an archived backup tape at Nuance, where the voice IP and software had ended up after various company acquisitions."

SPARE VOICE
Having spare voices allowed the team Jonathan Wood had put together in the UK, including Peter Benie at Cambridge University's Department of Engineering, to analyse one of them to find a new way to reproduce it. While the CallText 5010 circuit board had used both digital and analogue components to create the building blocks of Hawking's original voice, the hope was that software that could run on any computer could do the same job.

Various strategies were tried. One, by a company called Phonetic Arts among others, was to train a modern voice system to reproduce samples of Hawking's synthetic voice, where lines of recorded dialogue are put into a 'speech

OPPOSITE Synthesizer voice emulation prototype, developed to preserve Hawking's voice by his assistant Sam Blackburn.

library' and these sounds pieced together to generate new sentences. But, though these sounded convincing enough to the likes of Benie, Hawking did not agree. Further analysis by a speech technology expert, Patti Price, would tease out the differences that were obvious to Hawking.

Eventually Wood's team collaborated with Paweł Woźniak, an electronics engineering student in Huddersfield familiar with the increasingly rare hardware boards that Hawking depended on, Mark Green, an electrical engineer with Intel, and Jon Peatfield, a computer officer in Hawking's department (who sadly died during the project), to develop software capable of emulating his trademark voice. "Peter deserves the lion's share of the credit for writing the emulator," said Jonathan Wood, "but everyone's input was required for it to be a success."

On 26 January 2018, not long before Hawking died, Jonathan installed the new program onto his wheelchair. For the first time in more than three decades, Stephen could speak without having to depend on CallText. "I love it," he wrote.

Stephen Hawking's interaction with his computer would also evolve down the decades. In 1989, he could still move both hands and have a switch for each to control: one for his wheelchair computer and one for his desktop computer. But as his muscle control deteriorated, around 2004 he changed hands, now using his right hand to drive his chair.

But later that same year both hands were responding too slowly, so Hawking started to use an infrared switch mounted on his glasses which he would control by moving his cheek. The original device was off the shelf, and for years Hawking's assistant Sam Blackburn worked on adapting it; a later one was customized to his needs.

HANDS-FREE OPERATOR
As Stephen Hawking's condition worsened, his signals became harder to separate from random movements and noise. Fortunately, in 1997, he had met Gordon Moore, the co-founder of Intel, at a conference. Moore noticed that the computer Hawking used to communicate had an AMD processor and asked if he'd prefer a "real computer" with an Intel chip. From 1997, therefore, Intel sponsored and

> For the first time in more than three decades, Stephen could speak without having to depend on CallText. "I love it," he wrote.

provided Hawking's computer-based communication system and, around the time of his seventieth birthday, started work to provide a new interface, drawing on Jonathan Wood's experience of exactly how Hawking interacted with his computer.

Hawking himself described the process: "A cursor automatically scans across this keyboard by row or by column. I can select a character by moving my cheek to stop the cursor. My cheek movement is detected by an infrared switch that is mounted on my spectacles. This switch is my only interface with the computer." An Intel program called ACAT (Assistive Contextually Aware Toolkit), now available as open-source software, selected the letters, and Hawking's cheek switch was improved with digital sensors that wouldn't react to changes in ambient light, and modern signal-processing techniques to recognize his ever-smaller cheek movements.

ACAT included a word prediction algorithm provided by SwiftKey, a Microsoft program which had been trained on Hawking's books and lectures: "I usually only have to type the first couple of characters before I can select the whole word," he said. "When I have built up a sentence, I can send it to my speech synthesizer." Through ACAT he could also control the mouse in Windows to operate his whole computer: "I can check my email using Microsoft Outlook, surf the Internet using Firefox, or write lectures using Microsoft Word. My latest computer from Intel also contains a webcam which I use with Skype to keep in touch with my friends. I can express a lot through my facial expressions to those who know me well."

SYNTHESIZING LECTURES

Hawking could use the system to give his lectures too: "I write the lecture beforehand, then save it to disk. I can then use a part of the ACAT software called Lecture Manager to send it to the speech synthesizer a paragraph at a time. It works quite well, and I can try out the lecture and polish it before I give it." He did not actually have to be there to give his lectures, and as his mobility reduced, he would occasionally deliver them remotely, even once using "hologram" technology, but there was something immensely powerful and affecting having him in person before you.

As Hawking's physical condition continued to deteriorate in his later years, he investigated new assistive technologies, including, he told one interviewer, "eye tracking and brain-controlled interfaces to communicate with my computer". He also used a mute button to allow him to turn off his speech synthesizer while he was travelling, in case the judder caused his cheek switch to generate random output. He did not use the mute at mealtimes, though, recalled Jonathan Wood, even if it meant uttering random words, as he found it more entertaining. "Muting himself would have meant him being silent for the whole mealtime, which was a less preferable option."

SEE ALSO:
————

"There was something so iconic, dishevelled and cool – effortless, as if he is not trying," Redmayne told me. "There is an amazingly sexy quality of being confident in his own skin."

How Eddie Redmayne Got Stephen Hawking to Lend Him His Voice

Thanks to Hollywood, there was a second person to master Hawking's voice: the actor Eddie Redmayne, who depicted him in the Stephen Hawking biopic *The Theory of Everything*, which was based on a 2007 memoir by Hawking's first wife Jane. In his foreword to *Brief Answers to the Big Questions*, Stephen's final book, Redmayne recounted their first meeting, when he was "struck by his extraordinary power and his vulnerability".

HAWKING AS JAMES DEAN

Redmayne told me how he had spent months studying material about Hawking, from books to video, and came to think the young Stephen had something of James Dean about him. "There was something so iconic, dishevelled and cool – effortless, as if he is not trying," he told me. "There is an amazingly sexy quality of being confident in his own skin." One key insight had emerged when he talked to Hawking's son, Tim: "We were all respectful about the disease, but Tim said, 'Yes, we did get in Dad's wheelchair and use it as a go-kart. And we did put swear words in his voice machine.'"

Redmayne also worked with the Motor Neurone Disease Association and a neurology clinic at University College London, meeting some thirty patients. With a dance teacher he worked on the change in Hawking's movements as the disease took hold, and he wore prosthetics to complete a physical transformation that won him multiple awards, including an Oscar.

The magic key that allowed Redmayne to unlock Hawking's character materialized when, just days before filming began, he met the man himself to get the measure of him. Hawking takes an age to type into his voice synthesizer, so Redmayne only had a few sentences to go on, but the gravitational pull of his personality on those around him, and how he flirted with his co-star Felicity Jones, left Redmayne in no doubt of Hawking's charisma. Overall, Redmayne was struck by "his extraordinary force of personality, incisive wit, humour, and mischief – I describe it as a 'lord of misrule' quality. That was worth its weight in gold."

Redmayne was nervous about what Hawking would make of *The Theory of Everything*, but in his trademark tones Hawking had told him, "I will let you know what I think, good or otherwise." Redmayne need not have worried. Jonathan Wood was with him when he first saw the film at a private showing in London. "It was really emotional, even uncomfortable at times, as we could tell Stephen was crying through some of the scenes. I remember he was fully engrossed in it, and afterwards he said that at times he thought Eddie Redmayne was him."

After the screening, Hawking was so pleased with the movie he allowed the filmmakers to swap the synthetic voice they had created for his own, copyright version. "He offered us his voice," said Redmayne. "For me, that was the most wonderful thing." At the London premiere, the two of them posed happily together for the cameras, and Hawking celebrated with the film's director James Marsh by sipping champagne from a spoon.

ABOVE Eddie Redmayne and Stephen Hawking attend the after party for the EE British Academy Film Awards in 2015.

ABOVE One of the spoons from which Hawking liked to drink – he particularly enjoyed drinking tea.

Me and My Spoon

The British satirical magazine *Private Eye* has a long-running parody of vapid celebrity lifestyle columns, entitled "Me and My Spoon". That the magazine had featured as a stake in one of Stephen Hawking's scientific bets suggested that he was himself an avid reader.

"I always wish they'd done my dad," remarked Stephen Hawking's daughter Lucy. "He's probably the only person who actually cared about his spoons. Because my father needed assistance with eating and drinking, he was very fussy about having his own spoons." Hawking required a lot of fluid, and he particularly loved tea. "After a while he couldn't drink unaided from a cup," recalled Lucy. "So the carer would use a spoon to help him eat and drink. And there were particular spoons where he liked the dimensions, and they just worked for him." Out of all of them, one spoon was a little deeper than normal, allowing it to contain more tea. Even if he was in a Michelin-starred restaurant, Stephen Hawking would insist on using his own spoon. "I would say the spoon is probably the most significant innocuous object in his office," said Lucy, adding that his collection was "quite special".

ABOVE Bowl used for crushing medical pills, from the small kitchen within Hawking's office.

Even if he was in a Michelin-starred restaurant, Stephen Hawking would insist on using his own spoon.

F+B+T

Hawking's penchant for tea, which "he drank all day long", went beyond mere liking. He kept abreast of the medical literature ("he was always pleased by newspaper articles about the health benefits of tea!"), and believed that vitamin B and folic acid could help to reduce levels of homocysteine, an amino acid, in his body. Elevated levels of it have been postulated as harmful to both heart and brain. "He would take the most astonishing number of vitamin B and folic acid supplements per day with his tea," said Lucy. Bowls like this one were used to crush up the pills, which were kept in his office in travel bags (our

conservator Ruth Nightingale found some of these pills in one of his travel carrier bags, so they are now part of the collection). On his computer screen at routine intervals during the day would appear the gnomic instruction, "F+B+T". "Everybody would know what that meant," recalled Lucy. "He wants his folic acid. He wants his vitamin B. And he wants a cup of tea."

SEE ALSO:

Caffeinated Science, p.98

Caffeinated Science

Most of us need a cup of coffee to kick-start our thoughts at the beginning of a working day, and probably a cup of tea later on to perk us up once again. These mundane objects are testament to how, in this respect, Stephen Hawking was no different.

His studies thrived on long hours together with his fellow scientists in front of blackboards and computer screens, fuelled by caffeine, a neurotoxin made by plants to deter insect pests that in humans alters the biochemistry of the brain to make a person more focused and alert.

AN ENGLISHMAN AND HIS TEA

"Stephen liked strong black tea, for the taste, but also", according to his PA Judith Croasdell, citing his carers of many years standing, Viv Richer and Patricia Dowdy, "because his father had said drinking black tea was good for his digestive system."

Undoubtedly his favourite drink was tea, Croasdell added, though "Patricia often had to grind the coffee beans for his breakfast coffee."

for the rest of the day he loved to stop for tea, hence all the good-quality China cups and saucers, both in the office and at home. Stephen loved plain black Assam tea best – tea leaves, not teabags, made in a proper teapot. He also had a special tea strainer. The carers always doled his tea out of a China cup and saucer, using a deep teaspoon almost like a soup spoon.

Croasdell stressed the importance of getting the right kind of tea. "In 2001, at Heathrow Airport, Patricia Dowdy

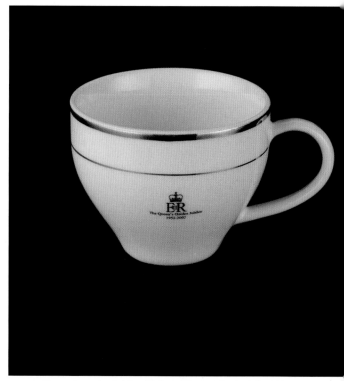

ABOVE Teacup from the Golden Jubilee of Queen Elizabeth II, 2002.

recalled Stephen once holding up a British Airways flight going from London to San Francisco in order to have a cup of his favourite black tea." She added, "He also insisted on having it in one of his China cups, which was always carried around in his black bag."

In 2001, at Heathrow Airport, Patricia Dowdy recalled Stephen once holding up a British Airways flight going from London to San Francisco in order to have a cup of his favourite black tea.

Sean Carroll, a professor of physics at Caltech, recounted how, after arriving at the airport, Stephen and his entourage first went shopping for groceries before having dinner. "While parked at a local supermarket, a stream of nurses travelled between the van and the store, bringing various samples of teabags for Stephen to choose from. Stephen, solid Englishman that he was, was very particular about his tea."

THE POTTER ROOM

There is plenty of evidence that tea propelled Hawking's science. Alan Yuille, Bloomberg Distinguished Professor of Cognitive Science and Computer Science at Johns Hopkins University, who did his PhD with Stephen Hawking on quantum gravity in the 1970s, was told that "Stephen believes the best work is done in the tea room."

At the Department of Applied Mathematics and Theoretical Physics at the University of Cambridge was a common room, known as the Potter Room, where daily teatime gatherings were held, following the tea-room tradition of the department's previous Silver Street building, which Stephen loved to attend. "At morning and afternoon tea his group would sit together in the DAMTP tea room," recalled Yuille, "and have informal discussions including physics and almost everything else."

"Our working life centred around the tea room," added Bruce Allen, another former student, now director of the Max Planck Institute for Gravitational Physics in Hanover, Germany.

Hawking's love of tea was acknowledged in a gift from Wu Zhongchao, one of his doctoral students, who went on to translate Hawking's books into Chinese and helped organize his trips to China. In 2004, he gave Hawking a magnificent Chinese tea jar, made of compressed tea, from Yunnan province. "You can break the jar and make drinkable tea from the shards," said Croasdell.

To meet the demand for caffeine for the rest of the time, Hawking's office was fitted with a kitchen counter under his bookcases, and included a refrigerator, microwave oven, sink, and storage space for essential equipment to brew tea and coffee.

Appropriately enough, one sign of Hawking's enduring influence is that, even today, one can buy tea and coffee cups adorned with his image, along with his words of wisdom.

SEE ALSO:

Me and My Spoon, p.96
Hawking's Living Legacy, p.217

TOP The tea room in the old Silver Street site of the Department for Applied Mathematics and Theoretical Physics. **ABOVE** Hawking's tin of tea.

The Anthropic Principle

Although removed from his office years before the acquisition of its contents by the Science Museum Group, there once hung in Hawking's office a sign declaring "Yes, I AM the Centre of the Universe." As with so many of the apparently straightforward objects in his office, there was more to this jokey and solipsistic sign than first met the eye. It goes to the heart of an issue that Stephen Hawking wrestled with for decades, including in his paper "The Cosmological Constant and the Weak Anthropic Principle" from 1982. The sign's words seem at odds with what scientists call the Copernican principle, which says that there is nothing special about the position or view of humans (even Stephen Hawking) in the universe. This contradicted the ancient Ptolemaic system, which did indeed put Earth at the centre of the universe.

These principles have powerful consequences for how we view our place in the universe. Just as Copernicus showed that we are not at the centre of the Solar System, so the American scientist Harlow Shapley concluded that our Solar System is not at the centre of our galaxy. Indeed, we now know that there are several hundred billion galaxies in the observable universe; many scientists now think our universe is one among many, part of what is called the multiverse.

Yet for nearly half a century, scientists have marvelled at how fundamental aspects of the cosmos – the laws of nature and around three dozen parameters that specify physical aspects of the universe – seem fine-tuned to an extraordinary degree to allow for the emergence of living things. As Stephen Hawking remarked in *A Brief History of Time*,

> The laws of science, as we know them at present, contain many fundamental numbers, like the size of the electric charge of the electron and the ratio of the masses of the proton and the electron ... The remarkable fact is that the values of these numbers seem to have been very finely adjusted to make possible the development of life.

> If the rate of expansion one second after the Big Bang had been smaller by even one part in a hundred thousand million million, the universe would have recollapsed before it ever reached its present size.

"WEAK" OR "STRONG"?

In 1973, in the light of these anthropic coincidences, the phrase "anthropic principle" took root at a Kraków symposium honouring Copernicus's five-hundredth birthday. At the meeting, the Australian-born English physicist Brandon Carter, a post-doctoral researcher in astrophysics at the University of Cambridge (and a contemporary of Martin Rees, who was just a couple of years behind Hawking in his studies), proposed the concept of the "weak" anthropic principle. Simply put, Carter pointed out that if our universe were not hospitable to life, then we would not be here to wonder about it.

This should be distinguished, he said, from a "strong" anthropic principle. Because there's nothing special or privileged about Earth or humanity (and because Stephen Hawking is not at the centre of the universe), we must live in a universe capable of supporting life, and only life-supporting universes are possible. To put it another way, the weak anthropic principle is tantamount to saying our existence may have been just luck, while the strong version rests on scientific theories that suggest our existence was likely to occur, perhaps even inevitable.

The most fashionable of the many ways to formulate the strong anthropic principle is in terms of the multiverse – that if many, possible universes exist, each defined by a different permutation of physical constants, life would have to arise in at least one.

Just as the creatures around us reflect the outcomes of countless chance events over the four billion years that life has thrived on our planet, so accidents during the earliest evolution of each multiverse universe, when it was tiny and the uncertainty of quantum mechanics ruled, were subsequently frozen into its physical blueprint. Even tiny alterations in the life-enabling constants of fundamental physics in this hypothesized multiverse could, as Stephen Hawking put it, "give rise to universes that, although they might be very beautiful, would contain no one able to wonder at that beauty". By the same token, however, it is inevitable that life will occur somewhere in the multiverse; that is, the multiverse appears to build in the strong anthropic principle.

But by the time he met with his former student, Thomas Hertog, in June 1998, Stephen Hawking had changed his mind. Now he declared that the anthropic principle is "a counsel of despair ... a negation of our hopes of understanding the underlying order of the universe."

In his book, *On the Origin of Time: Stephen Hawking's Final Theory*, Hertog recorded his surprise, given that "the early Hawking had frequently flirted with the anthropic principle as part of the explanation for the universe." Is this why the "centre of the universe" sign had vanished by the time we collected Stephen Hawking's office for the nation?

Hawking had grown increasingly worried that relying on the anthropic principle for the existence of the multiverse made it an untestable unscientific idea. He was bothered by how the multiverse formulation of the anthropic principle still revolved around us when currently there is no experimental evidence to prove the multiverse is real, and we have no way of knowing, for example, if we live in a universe that is teeming with civilizations or one where life is relatively rare. So which anthropic/biophilic universe should be ours?

THE "TOP-DOWN" APPROACH

In response to this conundrum, from 2006 Stephen Hawking and Thomas Hertog set out to develop an alternative theory of the origin of the universe, based on what Stephen called the "Top-Down" approach. To understand what Stephen meant by "Top Down", Hertog explains that the usual framework for prediction in cosmology works "bottom up": that is forward in time. By this method, one assumes that the universe has a single history with a well-defined starting point, and attempts to predict how it will evolve in the future. This is a fundamentally "classical" way of thinking.

The alternative view Hertog adopted with Stephen Hawking arises from looking at the universe from a quantum perspective. The many peculiar-yet-fundamental aspects of quantum mechanics imply that the observer – through the act of measuring a system – selects one out of many possible histories the system could have had. This is why they referred to their quantum thinking as top-down cosmology: "We read the fundamentals of the history of the universe backwards, not from the Big Bang of creation but 'from the top down'," said Hertog.

Hawking and Hertog's top-down thinking combines evolution, boundary conditions and observership into a single, holistic scheme of prediction. "A tangible past and future emerge out of a haze of possibilities by means of a continual process of both questioning and observing," writes Hertog. "It is as if the act of observation today retroactively fixes the outcome of the Big Bang 'back then'." This, Hertog goes on, "embodies the view that down at the quantum level, the universe engineers its own biofriendliness. Life and the universe are in some way a mutual fit, according to the theory, because, in a deeper

sense, they come into existence together." As Hawking said, "With top-down we put us back in the centre." What he meant by "centre" is not the privileged anthropic principle position, explained Hertog, but the "more subtle, participatory element of the quantum act of observation".

Today, leading minds working on quantum gravity, the leading candidate for a "theory of everything", remain split on whether anthropic explanations should be acceptable in their final theory.

SEE ALSO:

A Brief History of Time, p.104
Universe on a Beach Ball, p.150

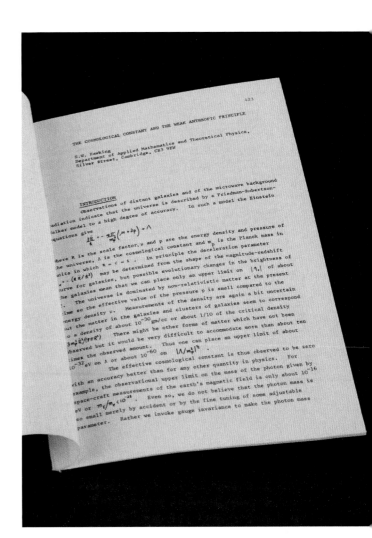

ABOVE Article on the Anthropic Principle from the scientific publication *Quantum Structure of Space and Time*, 1982.

Royal Recognition

A few years before the publication of *A Brief History of Time,* Stephen Hawking was appointed CBE, Commander of the Order of the British Empire, an honour which recognizes distinguished and innovative contributions. His award, announced in the New Year's Honours, was "for services to astronomical research" and reflected a growing public interest in black holes and the increasing academic recognition of his work, not least his election in 1979 to Lucasian Professor of Mathematics.

In 1989, he was also made a member of the prestigious Order of the Companions of Honour, a special award granted to those who have made a contribution of national importance over a long period. The members of this Order, of which there are 65 at any one time, currently include the actress Dame Maggie Smith, the artist David Hockney and former prime minister John Major.

DECLINED KNIGHTHOOD

However, in the late 1990s Hawking reportedly declined a knighthood. One oft-cited reason was that he objected to the UK's science funding policy; another was that he did not like titles.

Hawking did not seem to object to the monarchy, however, even though he did run over the toes of the future King Charles in 1977 as the then Duke of Cornwall was being inducted into the Royal Society. That was an accident and he was no republican: his office even contained teacups that celebrated the Queen's Golden Jubilee in 2002, and the fiftieth anniversary of the accession of Elizabeth II on 6 February 1952.

When the Queen came to open Cambridge's new Centre for Mathematical Sciences in 2005, no introduction was needed when her party visited Hawking's office. "Oh I know him!" the monarch declared.

The following May, he was a guest at the "Serving Beyond Sixty" reception that the Queen hosted at Buckingham Palace to celebrate the achievements of high-profile figures in their seventh decade and beyond, and returned a few months later for another reception, this time to celebrate the importance of British science at what was billed as a Science Day.

GOOD EVENING MA'AM

As well as celebrating science, Hawking was keen to use his encounters with royalty to raise the profile of disability issues, and in 2010 the Queen and Stephen Hawking were reunited, this time at the Chelsea Flower Show – Hawking famously quipping "Good evening, Ma'am" – in a garden commissioned by the Motor Neurone Disease Association and inspired by him, in which a small dark circular pond represented a black hole.

While hosting a reception at St James's Palace in 2014 to celebrate the work of the charity Leonard Cheshire Disability, the Queen asked Hawking whether he still had "that American voice". "Yes", Hawking replied, "it is copyrighted, actually".

SEE ALSO:

On Shoulders of Lucasian Giants, p.62
Picturing Hawking, p.73
"The Most Influential Disabled Person", p.206
Hawking's Last Wheelchair, p.208

OPPOSITE Commander of the Order of the British Empire (CBE), awarded to Stephen Hawking "for services to astronomical research", 1982.

No introduction was needed when the Queen's party came to Hawking's office. "Oh I know him!" the monarch declared.

A BRIEF HISTORY OF TIME

From the Big Bang to Black Holes

'This book marries a child's wonder to a genius's intellect. We journey into Hawking's universe, while marvelling at his mind'
Sunday Times

Introduction by Carl Sagan

STEPHEN HAWKING

A Brief History of Time

At its most fundamental, there are two kinds of popular science book: *A Brief History of Time* and "all the rest". I first heard that quip from another bestselling author named Steve (Jones, the snail geneticist) decades ago; the joke still holds true. Since it was published in 1988, *A Brief History of Time: From the Big Bang to Black Holes* has sold millions of copies. His (relatively) easy-to-digest mediation on cosmology, from the origins and structure to the eventual fate of the universe, turned Hawking into one of the greatest ever popularizers of science.

That breakout moment occurred in April 1988, when *A Brief History of Time* was first published and became a sensation. I can remember the transformation of Hawking into a global celebrity almost happening before my own eyes. He went from being the subject of a BBC *Horizon* programme in 1983, which gave a vivid glimpse of how he coped with disability in his research, to the recipient of a rock star reception at the book's launch in 1988, which I witnessed first-hand during a visit to the University of California, Berkeley, during his American tour.

The foreword of the first edition was written by Carl Sagan. This was a wise choice; Sagan was himself a household name because of his milestone 1980 TV series, *Cosmos*, which has since been seen by more than 500 million people. Sagan's foreword provided a vivid sense of Hawking's extraordinary combination of determination and brainpower when he described how, in 1974, he had witnessed the investiture of scientists into the Royal Society (the world's oldest continuously existing learned society, founded in 1660). Among them was a young man in a wheelchair who shook the world of physics that very year by proposing that black holes evaporate gradually. Sagan recalled how he had watched Stephen Hawking slowly sign his name in a book that also carried the signature of Isaac Newton, remarking, "Stephen Hawking was a legend even then."

NEWTON'S SUCCESSOR

Sagan's reference to Newton in the foreword of Hawking's book was highly significant: although separated by three centuries, both men sought to understand the workings of the universe. The presentation of Newton's *Principia Mathematica* to the Royal Society on 28 April 1686 marked a turning point in physics. At the start of Book III of his work, Newton proudly declared: "I now demonstrate the frame of the System of the World." But Stephen Hawking, in *A Brief History of Time*, was even more ambitious than Newton. He wrote of how he sought a "complete theory" and an answer to why the universe exists. "If we find the answer to that, it would be the ultimate triumph of human reason – for then we would know the mind of God."

In his foreword, Sagan adds that Hawking is the "worthy successor" to Newton and Paul Dirac, both former Lucasian Professors of Mathematics at the University of Cambridge, a role that Hawking took up in 1979. The reader is left in no doubt of the scientific superstar status of the book's author. Hawking himself liked to make this connection. In his book *The Universe in a Nutshell* he quipped: "Newton occupied the Lucasian chair at Cambridge that I now hold, though it wasn't electrically operated in his time."

But the reference to Newton is, in a sense, ironic. He wrote *Principia* in Latin and, unlike Hawking, was not at all

OPPOSITE Paperback edition of *A Brief History of Time: From the Big Bang to Black Holes*, published by Bantam Books in 1995.

Hawking sometimes began his lectures with the following remark, "I assume you all have read *A Brief History of Time* and understood it." This always got a big laugh.

interested in reaching a wide audience, only fellow scholars. Partly because of earlier disputes about his ideas and partly because he had come to believe that the weightier matters of truth were beyond the hoi polloi, Newton pitched *Principia* at advanced mathematicians. The same was not true for Stephen Hawking, who wanted to write a book that would be bought in airport bookshops paradoxically for the same reason that Newton held the opposite view: the questions both struggled with are fundamental to the universe.

A decade after his investiture into the Royal Society, Hawking approached Simon Mitton, editor of astronomy books at Cambridge University Press, with his ideas for a popular book on cosmology. Part of his motivation was financial ("I thought I might make a modest amount") and part an excitement at his discoveries that meant he "wanted to tell people about them".

Mitton read the first draft and was put off by the equations littered throughout. Eventually Hawking chose to go with the trade publisher Bantam rather than CUP, but in subsequent drafts he decided to drop all the equations but one: perhaps the most famous of all, Albert Einstein's $E=mc^2$ (energy is equal to mass multiplied by the speed of light squared, in other words energy and mass are different sides of the same coin).

THE LEAST-READ BOOK?

The book was undoubtedly popular in terms of sales, but there were limits to how far-reaching it can be said to have been. Hawking sometimes began his lectures with the following remark, "I assume you all have read *A Brief History of Time* and understood it." This always got a big laugh: despite being a worldwide bestseller, the joke was that people struggled to get through it. Jordan Ellenberg, a mathematics professor at the University of Wisconsin, even devised a way to measure this paradoxical effect – using data from Amazon – to create what he called the Hawking Index, a measure for how far people will read a book before giving up. It was said at the time that *A Brief History of Time* was the least-read book ever, though Ellenberg found that particular distinction went to *Capital in the Twenty-First Century* by the French economist Thomas Piketty.

Why did the public find *A Brief History of Time* hard going? For some fields of science, notably quantum physics, Hawking's signature metaphors, similes and

LEFT Foreign editions of *A Brief History of Time*. Hawking received hundreds of different editions from his publishers around the world, in a multiplicity of languages. These books were displayed in the Science Museum for the Hawking seventieth birthday exhibit.

ABOVE Priest Bode Wright reading Hawking's book in *The Simpsons*.

analogies are no substitute for the mathematics on which this difficult theory rests. Attempts to popularize quantum theory with words alone are about as satisfying as trying to summarize *War and Peace* with a haiku.

TIME AND POSITIVISM

When I interviewed Stephen Hawking in 2001, I asked him about imaginary time, a concept he discusses at length in the book: a reference not to the imagination but a mathematical idea, where an imaginary number is one that, when multiplied by itself, gives a negative result. He used it as a mathematical tool, but I wondered if he had a clearer picture of what imaginary time actually is?

He replied that any picture of time is a mathematical tool according to the positivist philosophy of science that he adopted:

In this, a physical theory is a mathematical model. We cannot ask if a model corresponds to reality, because we have no independent test of what reality is. All we can ask is whether the predictions of the model are confirmed by observation. Models of quantum theory use imaginary numbers, and imaginary time in a fundamental way. These models are confirmed by many observations. So imaginary

time is as real as anything else in physics. I just find it difficult to imagine.

So why was the book so extraordinarily popular? One common theory is articulated by Astronomer Royal Martin Rees. He knew Hawking well, having enrolled as a graduate student at Cambridge University in 1964 where he encountered a fellow student, "two years ahead of me, who was unsteady on his feet and spoke with great difficulty". According to Rees, *A Brief History of Time* is

not an especially good book (there are better popularizations of cosmology), but its success is due to the public fascination with Hawking the man, how on Earth he managed to write it, and the idea of his imprisoned mind that could still roam the entire universe. If he had been a world leading figure in genetics, I am not sure there would have been the same interest at all.

SEE ALSO:

Not the New Einstein?, p.66
The Anthropic Principle, p.100
The Cosmic Censorship Bet, p.128

ABOVE Expanded and updated tenth anniversary edition of *A Brief History of Time*, 1998.

A Brief History, in Brief

Such has been its lasting success and impact that *A Brief History of Time* almost warrants its own brief history. The book is still in print and selling over 30 years after its initial publication. In the 1996 edition of the book, and subsequent editions, more subjects were included, notably time travel and wormholes. Nevertheless, the main structure of the book remains unchanged.

The book begins by framing Stephen Hawking's research within the history of astronomical studies, notably the gradual development of the currently accepted "heliocentric" model of the Solar System, with the Sun at its centre. He details how experimental evidence from pioneering astronomical observations was put on firm mathematical foundations by Newton's *Principia* and considers how the topic of the origin of the universe (and time) was studied and debated over the centuries.

A key moment in our modern understanding of astronomy came in 1929, when American astronomer Edwin Hubble used the largest telescope of the time to reveal that the more distant a galaxy is from us, the faster it appears to be receding into space. Others concluded that the universe must be expanding, and a corollary of this expansion was that the universe was once immeasurably small and began with what we now call a Big Bang.

Hawking then moves on to consider Einstein's general theory of relativity, which overturned Newton's picture of gravity. Remember that apple falling from a tree said to have inspired Newton's theory of gravity? Well, according to the general theory of relativity an apple falls to the ground as a result of the warping of space-time, in contrast to Newton's view which described gravity as a force that Earth exerts on the apple.

QUANTUM MECHANICS

Relativity, in the guise of gravity, shapes and governs the universe at large scales. But at the very origins of the universe, when it was microscopic, it was instead subject to the rule of perhaps the strangest theory of all: quantum mechanics. In the book, Hawking describes its development in the 1920s by the German Werner Heisenberg, the Austrian physicist Erwin Schrödinger and the English physicist Paul Dirac.

By the fourth chapter, Hawking asks us to consider how Einstein's general theory of relativity can be reconciled with quantum theory in black holes and the Big Bang. He points out that at extremely high temperatures of the kind present at the moment of the Big Bang, various forces behave as one, and that theories which attempt to describe the behaviour of this "combined" force are called Grand Unified Theories.

Hawking's research was inexorably drawn to black holes, regions of highly warped space-time where extremely strong gravity prevents everything, including light, from escaping the event horizon, the black hole's boundary. Hawking first caught the attention of his peers in the late 1960s, working with Roger Penrose on how general relativity sometimes breaks down, resulting in what is called a singularity inside black holes – and, most probably, at the start of the universe. This implies that singularities mark the beginning and end of space and time, which was created during the Big Bang.

BLACK HOLES

In Chapter Seven, Stephen Hawking discusses his most remarkable idea. In 1974, he published a new theory which argued that black holes can "leak" radiation, slowly shrinking over time until they eventually "evaporate". The reason for this is down to one strange consequence of quantum theory: empty space isn't empty at all, as pairs of particles (one of matter and one of antimatter) are constantly popping into and out of existence. If they appear on the border of the event horizon – the point of no return from the gravity well of a black hole, as described by general relativity – they may find themselves on different sides of the horizon, with one sucked in, and the other escaping to form what became known as "Hawking radiation".

Even today, the result is regarded with awe by his peers. "Hawking's revolutionary discovery that black holes radiate was the first spectacular result in quantum gravity," remarked Nima Arkani-Hamed of the Institute for Advanced Study, Princeton, "suggesting a startling unification of space-time, quantum mechanics and thermodynamics that has set much of the agenda for fundamental physics."

As the title of his book suggests, time is a central topic, yet some of the temporal issues would be less familiar to a general reader. He discusses the arrow of time, and why "real time" (time as humans observe and experience it) seems to point from the past towards the future, and yet it is "imaginary time" which is inherent to the laws of science.

IMAGINARY TIME

This use of imaginary time emerged from a key idea that Hawking developed with his long-term collaborator, James Hartle, professor of physics at the University of California in Santa Barbara. Because the early universe is so small, it falls under the spell of the theory of quantum mechanics, where the use of imaginary numbers is common. "At one stroke we were working in quantum cosmology," Hartle recalled.

In quantum mechanics, a mathematical object called the wave function describes all the possible states of a quantum object and assigns to each of them a certain probability. Hawking and Hartle sought a wave function for the entire universe that describes all the possible universes that could have been – including ones in which our Solar System never formed, or in which life might have evolved in quite a different way. In 1983 this was a novel idea, as Hartle recalled:

> Much of my work after '83 was in explaining what that meant. The argument is simple: if the Universe is a quantum mechanical system it must have a quantum state. The state can't be discovered by observation. It therefore has to be posited and checked by its predictions.

They called it the "no-boundary proposal" because the wave function encompasses the entire past, present and future of the cosmos at once – doing away with the need for a blue touchpaper to be lit, seeds of creation or a creator. From this "no-boundary proposal" Hawking explained that the thermodynamic arrow of time emerges if and only if the universe is expanding.

And as we venture back towards the beginning of our universe, space and time become fuzzy, "and cap off, somewhat like the South Pole on the surface of the Earth". Time did not exist in the early universe. Stephen Hawking famously said that asking what came before the Big Bang "is like asking what lies south of the South Pole". According to Hartle, Hawking remarked a number of times over the years that "the no-boundary wave function of the universe that he and I worked out was the best thing that either of us had done."

After a foray through time travel, and his own "chronology protection conjecture", Hawking moves on to the unification of physics and the search for a theory of "quantum gravity" that is both internally consistent and explains observed phenomena just as well as (or better than) existing theories do.

By one view, popularized by Carl Sagan, "extraordinary claims require extraordinary evidence". However, in modern cosmology this relationship between empiricism and rational thought is not so straightforward. In 1983 it was proposed that the theory of inflation could be eternal, leading to a multiverse with an infinite number of bubbles, in which our universe is but one swelling bubble, the cosmic and physical properties vary locally from bubble to bubble, and where everything that can physically happen does happen, an infinite number of times. But if all these universes exist, they are separated from ours, unreachable and undetectable by any direct measurement (at least so far). This theory seems to lie beyond the reach of evidence and science.

The conclusion of *A Brief History of Time* returns to the efforts through human history to understand the universe and humanity's place in it, beginning with the insights of religion and philosophers, and moving to the modern era, when scientists have placed their faith in reason and empirical observation to accomplish "the ultimate triumph of human reasoning". At the time the book was written, "superstring theory" had emerged as the most popular theory of quantum gravity but, despite ongoing research, we are still waiting for a convincing version that can solve many of the currently unsolved problems in physics.

THE HEIR TO GALILEO, NEWTON AND EINSTEIN

The end of the book features three short biographies of Galileo, Newton and Einstein. From the books, memorabilia and pictures in Hawking's office, it is clear that these are his scientific heroes. Though he no doubt wanted to suggest that, as Newton famously said, "If I have seen further, it is by standing on the shoulders of Giants", the inclusion of these mini biographies may well have signalled to many readers that Hawking was the intellectual equal of the likes of Galileo, Newton and Einstein.

Stephen Hawking even used this quotation in the title of one of his other books, *On the Shoulders of Giants*, a thick tome on the great works of physics and astronomy that he himself edited. However, he concluded in his commentary that:

> Our understanding doesn't advance just by slow and steady building on previous work. Sometimes, as with Copernicus and Einstein, we have to make the intellectual leap to a new world picture. Maybe Newton should have said, "I used the shoulders of giants as a springboard."

2
Global
Celebrity

Extra-Terrestrial Encounters

Stephen Hawking's fascination with extra-terrestrial life extended far beyond this close encounter with *E.T.* and Steven Spielberg, the director, producer and screenwriter who had advised him on how to turn *A Brief History of Time* into a documentary film, which was released in 1991.

Since there ought to be many other stars whose planets have life on them, the greatest puzzle about alien encounters was, as Hawking remarked, "Why hasn't the Earth been visited, and even colonized?" This is known as the Fermi paradox, a reference to the Italian-American physicist Enrico Fermi who, during a casual conversation about UFOs with fellow physicists in the summer of 1950, famously asked, "Where is everybody?"

Hawking himself was careful to discount suggestions that UFOs contain beings from outer space, because he believed that "visits by aliens would be much more obvious, and probably also, much more unpleasant." There were various reasons why an alien encounter has not happened, he speculated: perhaps the probability of life spontaneously appearing is so low that Earth is the only planet in the galaxy, or indeed in the observable universe, in which it evolved. Or perhaps alien life did emerge, but it is mostly unintelligent and, as Hawking drily remarked, "it is not clear that intelligence has any long-term survival value. Bacteria, and other single cell organisms, will live on, if all other life on Earth is wiped out by our actions."

And that, naturally enough, led to the third reason why he thought aliens have not visited our Earth. "Intelligent life destroys itself. This would be a very pessimistic conclusion. I very much hope it isn't true."

Over the years, Hawking became increasingly alarmed about the potential for human-induced global catastrophe, and joined the Board of Sponsors of the Bulletin of Atomic Scientists, established in 1948 by Albert Einstein, who had himself been concerned by the growing prospect of a nuclear apocalpyse. When in 2007 the Bulletin moved

RIGHT Photograph of Stephen Hawking with director Steven Spielberg and his alien character E.T., taken around the time Spielberg advised him on how to turn *A Brief History of Time* into a documentary film, which was released in 1991.

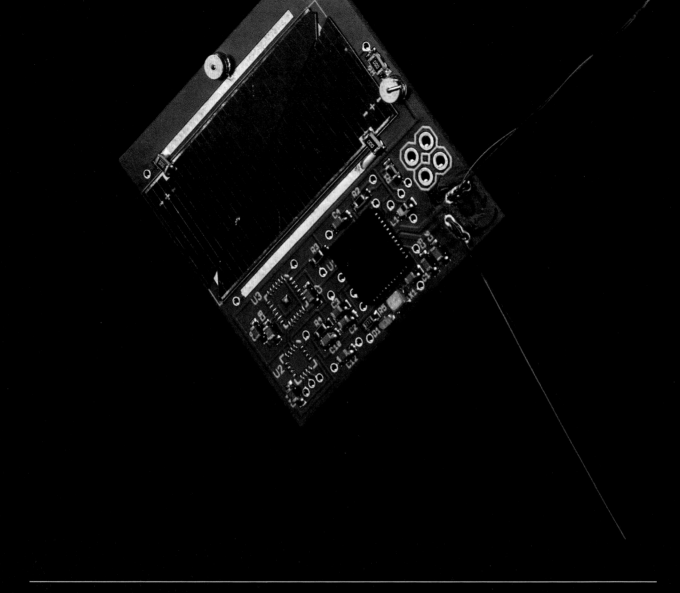

The greatest puzzle about alien encounters was, as Hawking remarked, "Why hasn't the Earth been visited, and even colonized?"

the minute hand of its famous Doomsday Clock closer to midnight, to warn the public how close we are to destroying our world with technologies of our own making, Hawking spoke out. "As citizens of the world", he said, "we have a duty to alert the public to the unnecessary risks that we live with every day, and to the perils we foresee if governments and societies do not take action now to render nuclear weapons obsolete and to prevent further climate change."

As for those elusive aliens, Hawking preferred a fourth possibility: that there are indeed other forms of intelligent life out there, but we have been overlooked by them, despite all the signals emerging from our planet, not least the signal bearing his voice that was aimed towards the nearest known black hole. And perhaps that is just as well.

DIDN'T TURN OUT SO WELL

In his later years, Hawking warned about the dangers of humankind encountering curious aliens. In one 2010 documentary series he suggested that alien civilizations sufficiently advanced to visit Earth may be hostile. "Such advanced aliens would perhaps become nomads, looking to conquer and colonize whatever planets they could reach," he said. "Who knows what the limits would be?"

And in the 2016 documentary *Stephen Hawking's Favourite Places*, he returned to his view that bumping into aliens would not turn out like Spielberg's *Close Encounters of the Third Kind*, or *E.T. the Extra-Terrestrial*: "Meeting an advanced civilization could be like Native Americans encountering Columbus. That didn't turn out so well."

Hawking was also interested in a backup plan if the Earth becomes inhospitable, whether due to our poor stewardship of the planet or hostile aliens. Among the items in Hawking's office is a Sprite, or ChipSat, a fully functional space probe built on a single circuit board, 3.5cm by 3.5cm, that weighs just 4 grams but can carry solar panels, computers, sensors and radios, as part of the Kicksat communications system.

THE STARSHOT MISSION

On 18 March 2019, over 100 of these tiny spacecraft were successfully deployed from the KickSat-2 satellite, with the first signals received the following day. This milestone was seen as one of the baby steps for a much more ambitious project to develop prototypes for interstellar travel called the Breakthrough Starshot mission, partially

funded by the Russian-born Israeli entrepreneur, venture capitalist and physicist Yuri Milner.

Stephen Hawking championed Starshot, and also served on the board of the $100 million project to launch fleets of tiny, sail-equipped probes to other star systems. The project followed in the wake of the alien-life-hunting $100 million Breakthrough Listen initiative, with which Hawking was also involved. "I strongly support ... the search for extra-terrestrial life," he declared. Presumably he wanted to know where a potential alien threat might lurk. Hawking thought that humankind would be well advised to keep the volume down on our intergalactic chatter, to discourage any "nomadic" aliens from becoming curious enough to pay Earth a visit.

If Starshot goes according to plan, powerful lasers will accelerate these "nanocraft" to about 20 per cent the speed of light, meaning they could get to Proxima Centauri, the nearest known star to our Sun and which hosts a potentially (just) habitable planet, a couple of decades after lift-off. "With light beams, light sails and the lightest spacecraft ever built," Stephen Hawking declared when Starshot was announced, "we can launch a mission to Alpha Centauri within a generation ... we are human. And our nature is to fly."

OPPOSITE Circuit board from a "KickSat" miniature swarm satellite, 2014–2017. **TOP RIGHT** Hawking with entrepreneur, venture capitalist and physicist Yuri Milner.

Stephen and Marilyn

On one wall of Stephen Hawking's office is a mocked-up photograph that shows Marilyn Monroe leaning against a Cadillac, with Hawking in his wheelchair beside her, looking as if they have just been on a date.

When Hawking had visitors, he liked to draw their attention to the image, a gift from the filmmaker Gordon Freedman, who in the early 1990s had worked on the documentary, *A Brief History of Time*. "Marilyn is an old girlfriend of mine," Hawking would like to say and, as if on cue, his guests would usually laugh. He also liked to joke that it was women who puzzled him most about the universe.

During Hawking's teenage years in the 1950s, Marilyn Monroe had become a cinema icon – and the object of global male interest – thanks to films such as *The Seven Year Itch* (1955), *Bus Stop* (1956), *Some Like It Hot* (1959) and *The Misfits* (1961). She died in 1962, at the age of 36, when Stephen Hawking was 20.

ABOVE Montage photograph of Stephen Hawking with Marilyn Monroe, created in 1992 as a gift from the filmmaker Gordon Freedman, who had worked on the documentary, *A Brief History of Time*.

ABOVE Mug depicting Marilyn Monroe, topped with her signature.

"Marilyn is an old girlfriend of mine," Hawking would like to say and, as if on cue, his guests would usually laugh.

MONROE MAD

Hawking's interest in her did not fade. When the TV presenter and journalist Piers Morgan asked him which three people he would choose to live with on a desert island for the rest of his life, Hawking first pointed out that he currently depended on a care team of eight and a technical assistant. But "if I were able-bodied", he added, "I would choose Marilyn Monroe, Einstein and Galileo."

Morgan, a former tabloid editor not prone to understatement, was struck by the posters and mug adorned with Monroe's image (a gift from his second wife) in Hawking's Cambridge office, along with a book of photographs of Marilyn. "He's Monroe mad," he declared.

Much of the Monroe memorabilia had been given to Hawking. "I very much enjoyed the film *Some Like It Hot*," Hawking told the *Los Angeles Times*, "so word got around that I admired Marilyn Monroe. My daughter and secretary gave me posters of her, my son gave me a Marilyn bag and my wife a Marilyn towel. I suppose you could say she was a model of the universe."

Monroe was indeed a constant presence in Hawking's life. When the documentary filmmaker Errol Morris was filming in a remarkable recreation of his Cambridge office,

they needed not only textbooks, computers and filing cabinets to dress the set, but also poster-sized photographs by Milton Greene and Philippe Halsman of Hawking's favourite pin-up, along with other Marilyn mementoes. During filming, one of the photos fell off the wall, prompting Hawking to type, "A fallen woman".

Morris responded with another quip: "You know, I finally understand what your attraction is to Marilyn Monroe. She was this very smart person who was appreciated more for her body than her mind."

A device employed by journalists was to ask the author of *A Brief History of Time*, "If you could go back in time, what would you do?" An encounter with Marilyn Monroe was, of course, a great reason for Hawking to make a temporal hop. "If I had a time machine, I'd visit Marilyn Monroe in her prime or drop in on Galileo as he turned his telescope to the heavens," he told one interviewer. Another asked him whether, if he could use a time machine, he would rather meet Marilyn Monroe or Isaac Newton? The answer was obvious: "Marilyn," he said, "Newton seems to have been an unpleasant character."

At Stephen Hawking's sixtieth birthday party, a Marilyn impersonator cut his cake and sang "I Want to Be Loved by You" – her most famous musical performance, from Billy Wilder's classic comedy *Some Like It Hot*.

Hawking's obsession with the actress would follow him to the grave. After his death in 2018, the Apollo astronaut Buzz Aldrin, who had once posed with him next to a cut-out of the movie star, wished him Godspeed and added, "I hope you're hanging out with Marilyn Monroe."

SEE ALSO:

Hawking's Last Wheelchair, p.208

> "If I had a time machine, I'd visit Marilyn Monroe in her prime or drop in on Galileo as he turned his telescope to the heavens."

ABOVE Framed poster of Marilyn Monroe. Knowing he was a fan, Hawking's daughter and secretary gave him posters of her, one of his sons gave him a Marilyn bag and his wife a Marilyn towel. **OPPOSITE** Illustrated biography of Marilyn Monroe, which featured over 200 photographs and 20 items of facsimile memorabilia from her personal papers.

MARILYN REMEMBERED
THE OFFICIAL TREASURES 1926-1962

Please Please Me

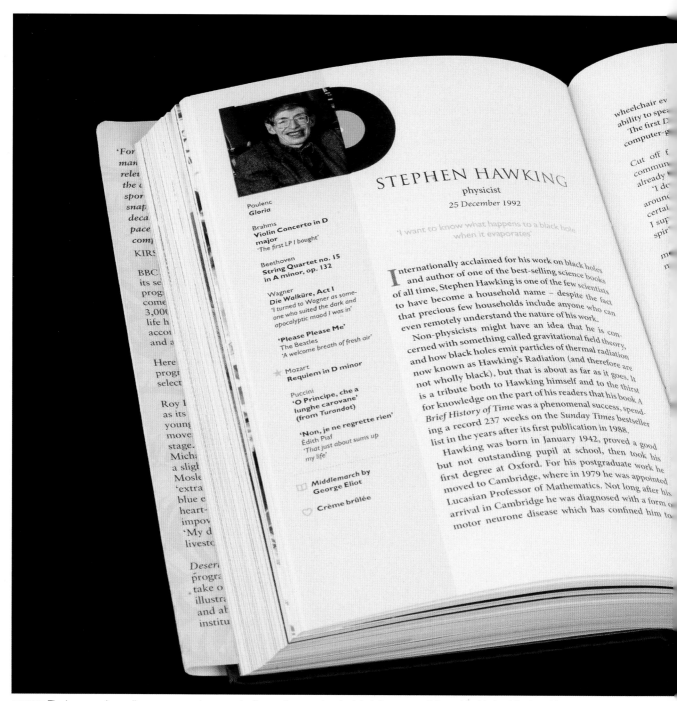

Poulenc
Gloria

Brahms
Violin Concerto in D major
'The first LP I bought'

Beethoven
String Quartet no. 15 in A minor, op. 132

Wagner
Die Walküre, Act I
'I turned to Wagner as some-one who suited the dark and apocalyptic mood I was in'

'Please Please Me'
The Beatles
'A welcome breath of fresh air'

Mozart
Requiem in D minor

Puccini
'O Principe, che a lunghe carovane' (from Turandot)

'Non, je ne regrette rien'
Édith Piaf
'That just about sums up my life'

Middlemarch by George Eliot

Crème brûlée

STEPHEN HAWKING
physicist
25 December 1992

'I want to know what happens to a black hole when it evaporates'

Internationally acclaimed for his work on black holes and author of one of the best-selling science books of all time, Stephen Hawking is one of the few scientists to have become a household name – despite the fact that precious few households include anyone who can even remotely understand the nature of his work.

Non-physicists might have an idea that he is concerned with something called gravitational field theory, and how black holes emit particles of thermal radiation now known as Hawking's Radiation (and therefore are not wholly black), but that is about as far as it goes. It is a tribute both to Hawking himself and to the thirst for knowledge on the part of his readers that his book *A Brief History of Time* was a phenomenal success, spending a record 237 weeks on the *Sunday Times* bestseller list in the years after its first publication in 1988.

Hawking was born in January 1942, proved a good but not outstanding pupil at school, then took his first degree at Oxford. For his postgraduate work he moved to Cambridge, where in 1979 he was appointed Lucasian Professor of Mathematics. Not long after his arrival in Cambridge he was diagnosed with a form of motor neurone disease which has confined him to

ABOVE The long-running radio programme where guests discuss the soundtrack of their lives was celebrated in this book by Sean Magee, *Desert Island Discs: 70 Years of Castaways.*

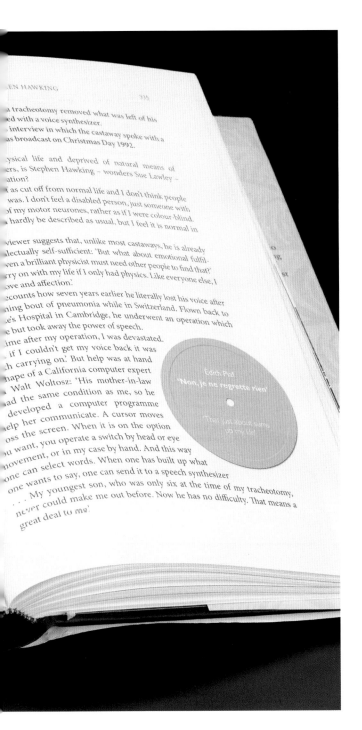

Lying flat on the shelves in Stephen Hawking's office was a 2012 book that celebrated the longevity of a radio classic: the seventieth anniversary of BBC Radio 4's *Desert Island Discs*. The show's format is simple: a guest is invited to choose eight musical tracks to take with them as they are cast away on a fictitious desert island – ideally ones associated with important memories or pivotal events in their life – as well as a single book and a single luxury.

In 1992 Stephen Hawking was the guest for a seasonal episode broadcast on Christmas Day. He revealed to the host, Sue Lawley, that the book he would take with him would be George Eliot's *Middlemarch*, and his one luxury crème brûlée, during a conversation that offered moving insights into how music and love had changed his life.

CLASSICAL MUSIC
Hawking had become aware of classical music as a fifteen-year-old. Long-playing records (LPs), which carried extended pieces of music on a single disc, had recently appeared in Britain. To play them, he upgraded an old wind-up gramophone and added a turntable and a three-valve amplifier. "At the time LPs were very expensive," he recalled,

> so I couldn't afford any of them on a schoolboy budget. But I bought Stravinsky's *Symphony of Psalms* because it was on sale as a ten-inch LP, which were being phased out. The record was rather scratched, but I fell in love with the third movement, which makes up more than half the symphony.

THE BEATLES
During his teenage years at the end of the 1950s, Hawking told Lawley, a period of relative wealth and stability had left some of the country's brightest young people feeling there was nothing much to aim for. "Most young people were disillusioned by what was called the Establishment," he recalled, and generally "bored with life". But the early 1960s had seen that apathy largely give way to the optimistic vibe reflected in one of Hawking's more unusual *Desert Island Discs* choices, The Beatles' "Please Please Me". Stephen Hawking was around the same age as Beatle Paul McCartney, and to him The Beatles were "light relief" and "a welcome breath of fresh air".

He revealed to the host, Sue Lawley, that the book he would take with him would be George Eliot's *Middlemarch*, and his one luxury, crème brûlée.

That record was released in 1963, at what would be a momentous time for Hawking. He had started his university education in October 1959 at the age of seventeen at University College at the University of Oxford, where he was lonely and bored, finding the undergraduate physics course "ridiculously easy". By his own admission, he did very little work, yet still graduated in 1962 with a first-class degree.

But by then he had also become increasingly clumsy, struggling with small tasks such as doing up his shoelaces, and his movements were erratic and ungainly. After an accident at a skating lake in St Albans, his mother took him to Guy's Hospital in London, where he was subjected to what he described as primitive and invasive medical tests. "A muscle sample was taken from my arm, I had electrodes stuck into me, radio-opaque fluid was injected into my spine."

The eventual diagnosis of motor neurone disease was only fully revealed to Hawking "years later", and it came as a shock. His musical tastes at what he remembered as "a confusing time" reflected, he explained to Lawley, the depths of his mood. "I discovered Wagner, who seemed to me to be as tragic a figure as I felt." However, he was by now a graduate student in Cambridge, and as the deterioration caused by his condition slowed, his diagnosis made him focus on his doctorate. The "prospect of an early death made me realize life was really worth living".

JANE WILDE

Another reason for his life becoming more tolerable, despite the increasing burden of his disease, was his budding relationship with Jane Wilde, a languages student whom he met through mutual college friends at a party in 1962. Getting engaged to her, he said, "lifted me out of the slough of despond I was in".

He would go on to marry Jane on 14 July 1965 in their shared home town of St Albans and she would provide a supportive home life for him and, to the surprise of the doctors treating him, the couple went on to have three children, Robert, Lucy and Tim. It was to Jane, he told Lawley, that he owed so much of his success in his career: "I certainly wouldn't have managed it without her. Jane looked after me single-handed as my condition got worse."

By that time, the doctor who had initially diagnosed Hawking had "washed his hands of me", and his father Frank, an expert on tropical diseases, became his primary source of medical support.

Along with The Beatles and Wagner, Hawking's other *Desert Island Discs* choices were mainly classical composers, perhaps more in keeping with what one might imagine would stimulate the mind of a genius: Brahms, Beethoven and Mozart. He also selected a soaring excerpt from Puccini's opera *Turandot*.

The pleasure that Stephen Hawking took from music throughout his life was matched only by his love of physics. No wonder, then, that in 2006 he was asked to nominate his three favourite classical works for a special concert at the Cambridge Music Festival themed around "Mozart, Music and Maths". As the University of Cambridge's then Lucasian Professor of Mathematics, Hawking was the obvious person to ask, and he selected Stravinsky's beloved *Symphony of Psalms*, along with Henryk Wieniawski's Violin Concerto No. 1 and Francis Poulenc's *Gloria*, which was also one of his Desert Island Discs.

NO REGRETS

Aside from The Beatles, there was one more exception to Stephen Hawking's classical choices as a castaway, one that he struggled to say because his voice synthesizer was, as he put it, "hopeless at French". That disc was "*Non, je ne regrette rien*" ("No, I don't regret anything"), composed in 1956 by Charles Dumont. Hawking settled on Edith Piaf's cover of it and this choice somehow summed up a grittier facet of his personality that was often overlooked: he possessed a relentless drive, a complete lack of self-pity and an unquenchable zest for life. To put it another way, as his daughter Lucy would trenchantly remark, he was "enormously stubborn". "That", he told Sue Lawley after they had listened to Edith Piaf's rousing version of one most popular French songs ever recorded, "just about sums up my life".

SEE ALSO:

The Butterfly Effect, p.196

ABOVE "Please Please Me" by The Beatles was one of the eight pieces of music Hawking selected for *Desert Island Discs*.

Stephen Hawking was around the same age as Beatle Paul McCartney, and to him The Beatles were "light relief" and "a welcome breath of fresh air".

To Boldly Go...

Stephen Hawking's celebrity was so great that it even extended beyond our world into science-fictional dimensions. In the entire history of *Star Trek*, the long-running American science fiction media franchise, he is the only scientist to ever play himself.

That seems somehow apt, given that the start of his scientific life began at the same time as the "Big Bang" of the *Star Trek* franchise: he submitted his doctoral thesis in October 1965, when he was 23 years old, the same year that "The Cage", the pilot episode of the American television series, was completed, though it was not broadcast on television in its entirety until 1988.

Stephen Hawking's appearance in the season six finale of *Star Trek: The Next Generation* resulted from a chance encounter in 1991, when he visited the Paramount Pictures back lot in Hollywood to introduce a documentary adaptation by Errol Morris of *A Brief History of Time*. The executive producer of the documentary, Gordon Freedman, recalled how Leonard Nimoy introduced Hawking at the screening. *Star Trek*'s executive producer, Rick Berman, happened to be there, discovered that Hawking was an avid Trekker and offered to give him a tour of the *Next Generation* set. "I told Berman that I was sure Stephen would appear on the show," said Freedman and, sure enough, they later learned he would do a cameo.

With the help of writer Ronald D. Moore, Hawking ended up with a role in "Descent", a double episode of *The Next Generation*, broadcast in 1993, that marked the twenty-sixth episode of the sixth season and the first of the seventh. "They told me, 'We need you to write a scene with Stephen Hawking,' recalled Moore, "and I'm like, 'Wait, *what?*'"

A holodeck simulation programmed by the android Data, Moore decided, would allow Hawking to play a game of poker with Albert Einstein and Isaac Newton. Berman sent Stephen Hawking the pages for the scene and "he punched the script up a little bit – he made a funny line a little bit funnier. It was delightful."

ENERGY AND MISCHIEF

Moore remembers how excited everyone on the set was when Hawking came in to do the filming. Only people involved in the scene – Hawking, along with Starfleet Lieutenant Commander Data (Brent Spiner, who by one account was nervous about the encounter), Sir Isaac Newton (John Neville) and Albert Einstein (Jim Norton) – were allowed in the room for filming. Everyone else was crammed out in the hallway trying to get a glimpse of the great physicist. During filming, Norton recalled, Stephen Hawking was "absolutely charming ... his eyes were full of life and vitality and energy and mischief."

In the broadcast episode, Hawking appears in the show's opening scene playing the card game on the holodeck, which, Data explains, is an experiment to reveal "how three of history's greatest minds would interact". This billing is a recurring theme throughout Hawking's numerous cameo appearances in series such as *The Simpsons*, *Futurama* and *The Big Bang Theory*. However, while Data hopes to witness meaningful conversations about humanity, the three scientists just poke fun at each other, and when a red alert is issued from the bridge he is forced to suspend the simulation.

In December 1998 my newspaper, the *Daily Telegraph*, reported on the *Star Trek* cameo, adding that Stephen Hawking would like to get more acting parts.

In his foreword to Lawrence M. Krauss's *The Physics of Star Trek*, Hawking mentions his delight at finding himself playing all scientific stars poker aboard the *Enterprise*:

A holodeck simulation programmed by the android Data, Moore decided, would allow Hawking to play a game of poker with Albert Einstein and Isaac Newton.

ABOVE Cast photo from *Star Trek: The Next Generation*, featuring Stephen Hawking with Jim Norton as Albert Einstein, and John Neville as Sir Isaac Newton, 1993.

Here was my chance to turn the tables on the two great men of gravity, particularly Einstein, who didn't believe in chance or in God playing dice. Unfortunately, I never collected my winnings because the game had to be abandoned on account of a red alert. I contacted Paramount Studios afterward to cash in my chips, but they didn't know the exchange rate.

(Hawking was owed 140 Federation credits.)

Hawking did tour the set at a later date and, upon seeing the warp core of the *Enterprise* in engineering,

commented, "I'm working on a space warp drive." And when he came to the transporter set, he quipped, "What a way to beat LA traffic."

SEE ALSO:
On the Shoulders of Lucasian Giants, p.62
Hawking on *The Simpsons*, p.142
America's Highest Honour, p.168
Will You Be My Valentine?, p.170
Hawking and *The Big Bang Theory*, p.178

The Cosmic Censorship Bet

At the centre of a black hole, according to work done by Roger Penrose and Stephen Hawking in the late 1960s, there must be a point of infinite density where the laws of physics break down.

Roger Penrose was troubled, because this had a curious consequence for determinism – the principle that, given the past and present, the physical laws of the universe must allow one, and only one, possible future. But if the laws of physics break down, and the singularity at the centre of a black hole seems to lie outside the domains of general relativity and quantum mechanics, then it would not be possible to predict the subsequent evolution of events there. In other words, if someone were to venture into one of these black holes, they could have an infinite number of possible futures, or no future at all.

For Penrose, however, determinism was sacrosanct. Perhaps the answer lay in a yet-to-be-developed theory of quantum gravity. To make the universe a tidy, easy-to-understand place, he proposed the Cosmic Censorship Conjecture, which says that all singularities formed by the collapse of stars or other bodies are kept safely isolated from the rest of the universe inside the "event horizon" around the resulting black holes. This event horizon is the boundary beyond which not only light cannot escape, but also events cannot affect an observer. As Stephen Hawking liked to paraphrase it: "God abhors a naked singularity."

During the early 1970s, Hawking's work had shifted through different fields, from cosmology to gravitational waves, and a deeper understanding of black holes. Hawking's student Gary Gibbons decided that, rather than focus on detecting gravitational waves, an increasingly fruitless quest at the time, he would look to find a way to challenge the Cosmic Censorship Conjecture. The simple answer was that he couldn't. Yet the conjecture was difficult to prove, said Gibbons.

A £100 BET

That did not discourage Hawking, at a meeting in Pasadena on 24 September 1991, from laying a bet in favour of cosmic censorship, arguing that "naked singularities are an anathema and should be prohibited by the laws of classical physics." The bet was against his Californian friends and scientific sparring partners John Preskill and Kip Thorne, who took the position that naked singularities are possible, "as quantum gravitational objects that might exist unclothed by horizons, for all the Universe to see". The stake was £100, and "the loser will reward the winner with clothing to cover the winner's

OPPOSITE Scientific bet with John Preskill and Kip Thorne, made in 1991. Preskill and Thorne took the position that naked singularities are possible, while Stephen Hawking bet that they are impossible.

To make the universe a tidy, easy-to-understand place, Penrose proposed the Cosmic Censorship Conjecture, which says that all singularities formed by the collapse of stars or other bodies are kept safely isolated from the rest of the universe inside the "event horizon" around the resulting black holes.

Whereas Stephen W. Hawking firmly believes that naked singularities are an anathema and should be prohibited by the laws of classical physics,

And whereas John Preskill and Kip Thorne regard naked singularities as quantum gravitational objects that might exist unclothed by horizons, for all the Universe to see,

Therefore Hawking offers, and Preskill/Thorne accept, a wager with odds of 100 pounds stirling to 50 pounds stirling, that when any form of classical matter or field that is incapable of becoming singular in flat spacetime is coupled to general relativity via the classical Einstein equations, the result can never be a naked singularity.

The loser will reward the winner with clothing to cover the winner's nakedness. The clothing is to be embroidered with a suitable concessionary message.

John P. Preskill Kip S. Thorne

Stephen W. Hawking   John P. Preskill & Kip S. Thorne
Pasadena, California, 24 September 1991

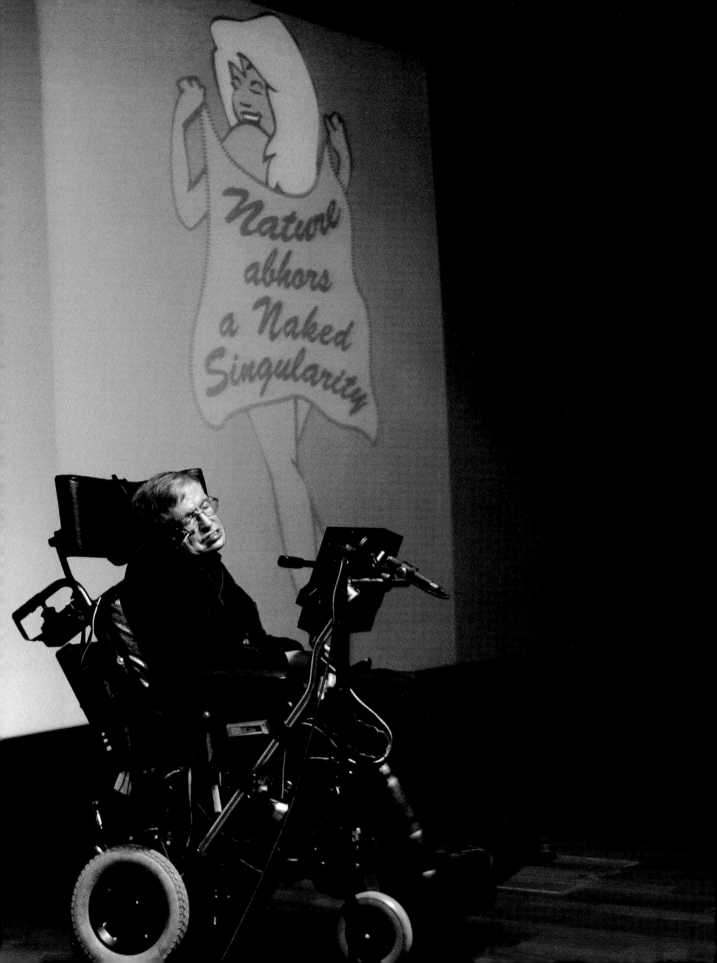

ABOVE Names of the three signatories on the Science Museum file holding the 1991 bet. **LEFT** Sir Roger Penrose, physicist, mathematician and philosopher of science. Penrose is Emeritus Rouse Ball Professor of Mathematics at the Mathematical Institute of the University of Oxford. **OPPOSITE** Hawking gives a lecture at the Bloomfield Museum of Science in Jerusalem, 2006. The screen behind shows the image he chose to put on the T-shirts he gave to Preskill and Thorne, despite having lost the bet.

nakedness. The clothing is to be embroidered with a suitable concessionary message."

Eventually, on 5 February 1997, Hawking conceded the bet "on a technicality", accepting that naked singularities can form under very special "non-generic" conditions. Kip Thorne recalled how they had originally offered to add the requirement to the bet that the solution be generic. "Stephen said no, he didn't need it. But it turned out that he did need it."

A few days later the *New York Times* reported on Hawking's lost bet: "A Bet on a Cosmic Scale, And a Concession, Sort Of", explaining how supercomputer calculations by Matthew Choptuik of the University of Texas in Austin suggested there could be special circumstances in which a naked singularity might be created from a collapsing black hole, either by nature or perhaps even by some advanced civilization. "The chance of this happening, Dr Choptuik said in an interview, would be comparable to standing a pencil upright on its sharpened tip."

For the "clothing … embroidered with a suitable concessionary message" Hawking gave Preskill and Thorne T-shirts emblazoned with a blond woman (was it Marilyn Monroe?) coyly adorned with a towel bearing the defiant legend, "Nature Abhors a Naked Singularity."

The loss did not dampen Hawking's appetite for wagers. Around the same time, the three physicists agreed to make another bet, this time on the question of whether information behind the event horizon of a black hole is irretrievably lost, or can be recovered from the radiation emitted as it evaporates. But while Hawking had been conservative on the issue of "naked singularities", he was heterodox when it came to what was known as the black hole information paradox.

On 5 February 1997, the three of them renewed the Cosmic Censorship Conjecture bet, on the grounds that "Stephen W. Hawking (having lost a previous bet on this subject by not demanding genericity) still firmly believes that naked singularities are an anathema and should be banned by the laws of classical physics." This time, the winner's clothing prize was "to be embroidered with a suitable, *truly* concessionary message".

Though Gary Gibbons had been studying the Cosmic Censorship Conjecture since the early 1970s, he acknowledges it remains "an open question", although when it comes to ordinary black holes the consensus is that nature does abhor naked singularities.

SEE ALSO:
A Brief History of Time, p.104

Whereas Stephen Hawking and Kip Thorne firmly believe that information swallowed by a black hole is forever hidden from the outside universe, and can never be revealed even as the black hole evaporates and completely disappears,

And whereas John Preskill firmly believes that a mechanism for the information to be released by the evaporating black hole must and will be found in the correct theory of quantum gravity,

Therefore Preskill offers, and Hawking/Thorne accept, a wager that:

When an initial pure quantum state undergoes gravitational collapse to form a black hole, the final state at the end of black hole evaporation will always be a pure quantum state.

The loser(s) will reward the winner(s) with an encyclopedia of the winner's choice, from which information can be recovered at will.

Stephen W. Hawking & Kip S. Thorne   John P. Preskill

Pasadena, California, 6 February 1997

Betting on the Black Hole Information Paradox

When Stephen Hawking began to stitch together the theories of the very big and very small, a problem emerged in the form of a paradox. This would inspire one of his famous bets with his peers.

Paradoxes, which frame puzzling contradictions, are of intense interest to physicists because they highlight flaws in current thinking and can help to reveal deeper theories of reality. Over many decades, Stephen Hawking would struggle with his "black hole information paradox", which resulted from a tension between the frameworks of quantum theory and general relativity. The former describes the subatomic realm as a domain of individual particles, or quanta, where there are sudden jumps, while general relativity depicts happenings on the grandest stage of all, the universe, in terms of smooth distortions and warps of space-time.

The paradox was a consequence of Stephen Hawking's milestone 1974 discovery that, by blending quantum mechanics and relativity, black holes are not entirely black, but glow: they give off heat, also introducing a third great branch of physics to the mix required to understand black holes – that of thermodynamics, the science of heat and work.

Black holes emit Hawking radiation and have a temperature and entropy, a key thermodynamic quantity which can be thought of as a measure of disorder and always increases. To understand the paradox that troubled Stephen Hawking, you need to know that entropy

is linked to information: the more disordered something is (or the higher the entropy), the more information we need to describe it. Imagine, for example, a string of 24 letters. If they are all the same, say X, then the progression is very ordered, low in entropy, and all we need is to say is "only Xs". But if the string is random, and high in entropy, then every letter has to be spelled out, and this requires more information.

No matter what you throw into a black hole, from a rocket to a pile of books of the same mass, its Hawking radiation will stay the same, as if black holes are oblivious to what they have consumed and all information about their diet has been erased. Classical general relativity says the matter is stretched and crushed and destroyed in the singularity that lurks within a black hole. Moreover, since the black hole evaporates as it gives off Hawking radiation it eventually disappears, along with all the information it once carried.

The black hole information paradox arises because this picture contradicts a key concept in quantum mechanics called unitarity, which, loosely speaking, says that anything that can happen can be undone; hence, information cannot be fundamentally lost. So, if the information did not leak out with the Hawking radiation, and the black hole evaporates, what happened to it?

GOD THROWS DICE WHERE THEY CAN'T BE SEEN
A famous bet between Stephen Hawking and his friends at Caltech, articulated in the document opposite that they all signed on 6 February 1997, was over how to resolve the black hole information paradox. Kip Thorne and Stephen Hawking championed the view that information is

OPPOSITE Bet with John Preskill and Kip Thorne on the black hole information paradox, 1997.

If the information did not leak out with the Hawking radiation, and the black hole evaporates, what happened to it?

Spaghetti vs Firewalls

If you fell headlong into a large black hole, you would not notice anything until you were well inside and had begun to be pulled lengthways and crushed sideways, a process called spaghettification, which emerges from a central principle of relativity (equivalence principle).

This idea was first challenged by Samuel Braunstein of the University of York, who in 2009 noted that around the black hole an "energetic curtain" would descend, of radiation formed from the sister particles of those given off by the black hole in the form of Hawking radiation. A poor astronaut would simply burn up. More importantly, said Braunstein, "it's curtains for the equivalence principle", though he thought that for general relativity there could be a way around this problem.

A paper published in 2012 suggested that quantum effects would indeed mean that anyone who tumbled into a black hole would be burned to a crisp. This became known as the "AMPS firewall", an initialism for the paper's authors: Joseph Polchinski, a string theorist at the Kavli Institute for Theoretical Physics in Santa Barbara, California; two of his students, Ahmed Almheiri and James Sully; and fellow string theorist Donald Marolf at the University of California, Santa Barbara.

Whether or not firewalls exist for any real astrophysical black hole is unclear, as Braunstein notes, since for any but the most microscopic black hole a firewall would take far longer than the age of the universe to form. "The process is far too slow."

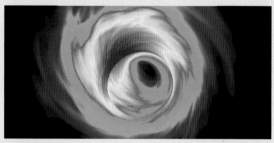

ABOVE An image from astrophysicist Scott Noble showing the entrance to a black hole. Colour represents the intensity of light emitted by gas circling the entrance; red is brightest and blue is the dimmest.

destroyed, arguing that the radiation from the black hole would be scrambled, so when it finally evaporated, any information about its contents would indeed be lost. "Not only does God play dice", as Hawking remarked, "but he sometimes confuses us by throwing them where they can't be seen". John Preskill, on the other hand, bet that in principle the information could be recovered, on the basis that, when physicists finally devise a theory of quantum gravity, we would understand how this could occur.

The winning side would receive an encyclopaedia of their choice, "from which information can be retrieved at will".

HOLOGRAPHIC UNIVERSE

Subsequent work by Gerard 't Hooft at Utrecht University, Leonard Susskind at Stanford University, Juan Maldacena at the Institute for Advanced Study in Princeton and several other string theorists suggested, Preskill recalled, "that information is encoded in black hole space-times in a very subtle way". This arose from their use of the concept of a holographic universe, in which any three-dimensional region of our universe can be described by information encoded on its two-dimensional boundary. Or, to put it in terms of the black hole information paradox, the information swallowed by a black hole can be encoded on its surface.

There was, however, a complication that is often ignored: they used what is called anti-de Sitter space, where de Sitter space is an approximation of our current universe. Anti-de Sitter space is by definition quite different, so, for example, any object thrown on a straight line will eventually return.

Even so, these arguments apparently helped sway Hawking. In the wake of work he presented at the Seventeenth International Conference on General Relativity and Gravitation, in Dublin on 21 July 2004, Hawking conceded the bet: he had come to believe that black hole horizons should indeed leak information, and presented Preskill with a copy of *Total Baseball: The Ultimate Baseball Encyclopaedia*. Implying that the information obtainable from a black hole was as useless as "burning an encyclopaedia", Hawking joked "but maybe I should just have given him the ashes". Thorne and others, however, remained unconvinced.

ABOVE Professor John Preskill raises a baseball encyclopaedia he won in a bet with Hawking that all matter does not disappear after entering a black hole, 2004.

SOFT HAIR

Although Hawking had accepted he had lost the black hole information paradox bet, in 2018 another insight into the information paradox emerged, when his final paper challenged his earlier thinking that the amount and temperature of the radiation emitted by a black hole depend only on the black hole's mass, spin and charge. In other words, whatever you throw into a black hole, once it is inside you will be unable to tell what has been swallowed. This was articulated in John Wheeler's memorable aphorism, "Black holes have no hair."

On the contrary, argued Stephen Hawking: it might be that quantum mechanics holds good and information about the collapse is encoded in some way within Hawking radiation. With his Cambridge colleague Malcolm Perry and Andrew Strominger of Harvard University, Hawking reported in his last paper in 2018 that no-hair theorems were not correct, and needed to take account of what became known as "soft hair". Stretching the hair analogy, his idea was (as I can personally attest) that if you look closely at bald people, you will find that their scalp is adorned with a fuzz of soft hair. The same is true of the event horizon of a black hole which, Hawking and his team argued, has more structure than thought – it also has "soft hair" that makes one black hole distinguishable from another, and allows information from the formation of the black hole to be preserved.

However, as Sam Braunstein of the University of York points out, it is "far from clear that the soft hair resolves the information paradox. For example, how is information encoded in a telephone directory that falls into the black hole 'bleached off' and converted into Hawking radiation?"

Even so, many physicists believe now that information escapes from inside a black hole, although the details and mechanism remain unclear. We still need to find a way to account for the entirety of the information apparently lost and, with other resolutions possible, the information paradox remains an active field of research in quantum gravity. As for Kip Thorne, though he thinks Stephen Hawking's "soft hair" paper is "intriguing", he never has conceded the bet: "I still think it likely that Stephen was correct originally: information is lost."

Thorne's reasons are twofold. First, he believes that backward time travel (through geometric structures called closed time-like curves) might be possible on the subatomic scales where quantum effects reign. Secondly, he uses a different formulation of quantum theory, devised by the Nobel Prize winner Richard Feynman, which is capable of dealing with closed time-like curves at the microscopic, quantum level, and predicts that they can "eat" information. "The Feynman approach, which was also used by Stephen's collaborator Jim Hartle, is the deeper approach, I suspect," says Thorne, "and it allows for information loss". To avoid information loss using conventional quantum theory, Thorne complains that his colleagues have to "undergo contortions", such as invoking firewalls.

SEE ALSO:
Hawking Radiation, p.46
Hawking in Space, p.164

The paradox was a consequence of Stephen Hawking's milestone 1974 discovery that, by blending quantum mechanics and relativity, black holes are not entirely black, but glow.

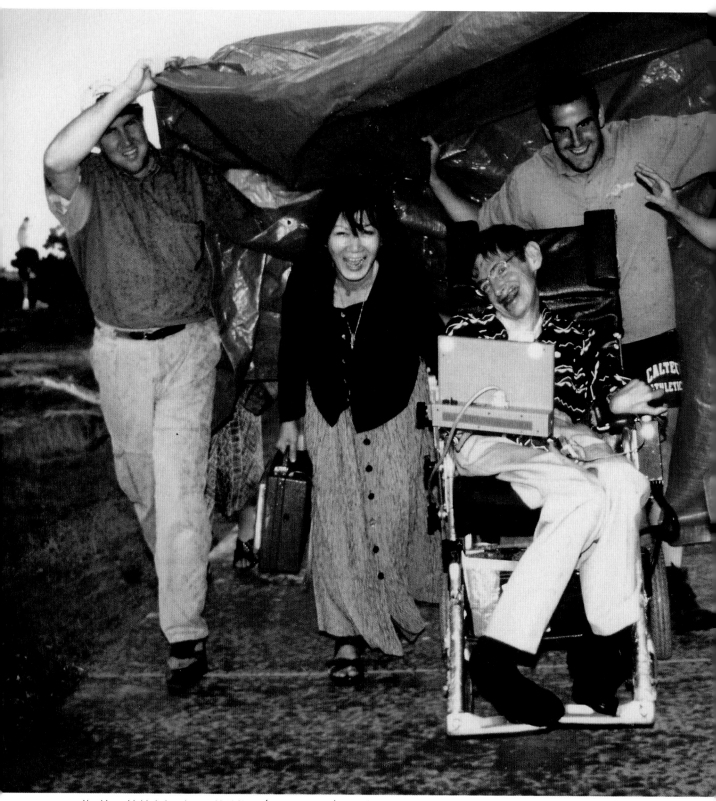

ABOVE Hawking with his beloved nurse Mooi Chong (second from left) sheltering from a rainstorm.

Unsung Heroes

An extraordinary mental toughness enabled Stephen Hawking to live a productive life despite his overwhelming disability, but he still relied on four full-time carers to help him with daily activities that most of us take for granted, and simply to keep him alive.

This is one his nurses, Mooi Chong (second from left), here sheltering from the rain with Hawking and students under a tarpaulin. "Stephen wanted this photo in his office as it showed one of his favourite nurses," commented his personal assistant, Judith Croasdell. "He's in his old wheelchair that was used as a spare on trips. My guess is that's probably California." Tragically, Mooi died of cancer in 2000. "He was extremely fond of her," said Croasdell. This is one of several pictures of her in his office.

LIFE ON THE BRINK

A glimpse of the importance of his carers came in Croasdell's essay, "Life on the Brink", one of many published in the German-language compilation *Stephen Hawking: Denken Ohne Grenzen* (Stephen Hawking: Thinking Without Limits). In it, she described how one young nurse had left her alone to look after the "Hawk", as Croasdell called him. "As soon as she left the room, his head lolled forward on his chest like that of a rag doll. As a result, his chin covered the breathing hole through which he drew air." Croasdell had to support his head for ten minutes so that he could breathe, until the nurse returned.

Hawking's own memoir also offers a sense of how precarious his life could be, acknowledging that his second wife Elaine's being a nurse "saved my life on several occasions".

His first wife Jane had refused to allow doctors in Switzerland to switch off his life support, after he had contracted pneumonia, and in her autobiography she gives a vivid account of organizing nursing care during the weeks of intensive care which followed, writing that "we floundered in an endless state of crisis."

SEE ALSO:

Hawking's Voice, p.90

ABOVE Photograph of Stephen Hawking with American President Bill Clinton, First Lady Hillary Rodham Clinton and Stephen's then wife Elaine Mason, during a visit to the White House in March 1998.

Presidents and Philanthropists

When Stephen Hawking visited the White House in March 1998, he met a fellow graduate of University College, Oxford, along with a philanthropist who would become a valued supporter and friend.

The then US president Bill Clinton and his wife Hillary had invited Hawking to deliver a Millennium Lecture, as part of a series "to highlight the creativity and inventiveness of the American people".

Their shared Oxford connection was almost the first thing Stephen Hawking mentioned when he arrived at the White House with his then wife, Elaine Mason. A decade after Hawking was an undergraduate at University College, Clinton had arrived in October 1968, aged 22, as one of 32 American Rhodes scholars. "I don't know how you found the food, but in my time it hadn't improved since King Alfred had burnt the cakes," said Hawking.

At the end of the talk President Clinton remarked that, while technology had enabled Hawking to speak, "it is also true, in my mind, that he is a genuine living miracle because of the power of the heart and the spirit."

ABOVE Framed photograph of Stephen Hawking with the American philanthropist Dennis Avery.

The President later joked that he was relieved his distinguished visitor had complained about the food, rather than ask "some incredible comparative academic question about our experiences there".

MILLENNIUM EVENINGS
Hawking's lecture was the second in a series of White House Millennium Evenings. Held in the East Room of the White House, he discussed "Imagination and Change: Science in the Next Millennium", beginning with his cameo in *Star Trek* and ending with a prediction that "the human race needs to improve its mental and physical qualities if it is to deal with the increasingly complex world around it ... [and] meet new challenges like space travel ... if biological systems are to keep ahead of electronic ones." At the end of the talk President Clinton remarked that, while technology had enabled Hawking to speak, "it is also true, in my mind, that he is a genuine living miracle because of the power of the heart and the spirit."

THE AVERYS
During his visit, Hawking also met the philanthropist Dennis Avery, who had studied at Cambridge, and was "astonished by his interest in the mysteries of the cosmos and its origin". Married to Sally Tsui Wong-Avery, a fellow lawyer, Avery supported charitable causes worldwide, and had already been told about Stephen by his father, Stanton Avery, who chaired the Board at Caltech, which Stephen had regularly visited since the 1970s.

Their support for Stephen and the University of Cambridge began in 1998, the same year they met, and led to the foundation of the university's Centre for Theoretical Cosmology. The Stephen W. Hawking Chair of Cosmology was also set up by Sally Wong-Avery, and whoever is selected for it works in Stephen's office.

The Averys were in 2009 awarded the Chancellor's 800th Anniversary Medal for Outstanding Philanthropy by the university at a ceremony at Buckingham Palace. And the same year they were Stephen's honoured guests when he was awarded the Presidential Medal of Freedom by President Obama.

Dennis Avery died in 2012, and Stephen kept three framed photographs in his office illustrating their long relationship and how they, and Hawking's wheelchairs, had changed over the years. In a tribute, Stephen said that Dennis had been to his home on many occasions, and that he stayed on to help wash up the dishes, which "tells you something more than words can express".

In 2013, Stephen was well enough to travel to Kearny Mesa in San Diego to help remember the late philanthropist at the Chinese Bilingual Preschool, which had been founded by the Averys. Officials planted a peach tree, a symbol of longevity in China.

Scientist-Rock Star

Stephen Hawking was often referred to as a rock star of science, which may sound a cliché, but in his case had some basis. His voice has featured on tracks by world-renowned artists such as Pink Floyd, Orbital and Vangelis.

A book of photographs by Anton Corbijn, the world's most famous photographer of rock stars, gives a nod towards Stephen's unique status. Taken in 1999, in black and white, the image of Hawking is dominated by the reflection in his teardrop sunglasses. "It is a stunning photo," said his daughter Lucy, "and marks a growing change in perception of scientists in popular culture from white-lab-coat-wearing nerds to a cooler, more iconic style of presentation."

Another pair of sunglasses dominates a later photograph, framed and prominently displayed in Hawking's office. The portrait, by Jaime Travezan for a magazine, shows Hawking sitting in front of a poster for the 2013 documentary on him, a copy of which was in his office. The poster was used, like others on hand, to hide the kitchen behind his office, and other distracting features, whenever a film crew or photographer joined him for a shoot.

Stephen seemed to enjoy the experience with Travezan, and posted several pictures from the shoot on his Facebook profile. "He did love showing off," remarked Lucy, with the faintest of sighs.

ABOVE A pair of Hawking's sunglasses with infrared cheek movement sensor for communication, 2012–2015.

OPPOSITE Framed photograph of Hawking wearing sunglasses in front of *Hawking* film poster, 2014–2015.

A REMARKABLE MAN
A REMARKABLE STORY

HAWKING

"AN ENGAGING PORTRAIT"

Hawking on *The Simpsons*

Stephen Hawking's real-world wheelchair was customized for his needs, but here, in this *Simpsons'* World of Springfield interactive figure, there are some additional, handy-but-fanciful modifications, including helicopter rotors and a spring-loaded boxing glove. "Helicopter blades would be very useful," Hawking later joked. "I don't use a boxing glove," he added – "though sometimes I'm sorely tempted".

The action figure, based on Hawking's first guest appearance as himself in *The Simpsons*, was created in 2003, with the figurine's computer screen reading, "If you're looking for trouble, you've found it." The line came from "They Saved Lisa's Brain", written by Matt Selman and directed by Pete Michels, which was the penultimate episode of the tenth season of the American animated television series and first aired on the Fox network in the United States on 9 May 1999.

"THEY SAVED LISA'S BRAIN"
The episode revolves around the Springfield chapter of the high-IQ society Mensa, which takes control of the town, hoping to improve the lives of Springfieldians through the rule of the smartest. As with many *Simpsons'* plots over the years, the idea of a ruling class comprising only the most intelligent is a provocative one, which raises some profound questions.

The science behind capturing and measuring intelligence certainly seems far more complex than the single, simple IQ score as used by Mensa. My own research on mass cognitive testing with colleagues in Canada (published in the journal *Neuron* in 2012) concluded that intelligence has different components that are anatomically distinct within the brain, and cannot be boiled down to one single number. Hawking himself paid little heed to IQ scores: a 2004 *New York Times* article quoted him as saying that "People who boast about their IQ scores are losers."

Nevertheless, for the plot to work, the *Simpsons'* writers needed someone who would be perceived as smarter than all of Springfield's Mensa members. The showrunner, or producer, Al Jean, who had studied mathematics at Harvard, thought Hawking would be ideal, especially having heard that he was a fan of the show.

It was certainly not the only time Jean brought his own interest in mathematical physics into the show. At a Science Museum event in 2014, he shed light on how and why it has regularly embellished episodes of *The Simpsons* (along with its sister show *Futurama*), with references to degree-level mathematics such as Fermat's Last Theorem or mathematically interesting numbers like a "Mersenne prime", a "perfect number" and even a "narcissistic number". "Everything else in showbiz is narcissistic," reflected Jean. "Why not numbers?"

HAWKING AND HOMER
Hawking recorded his performance for "They Saved Lisa's Brain" in the first week of December 1998, but by one account nearly missed the chance, as before his flight to Los Angeles, where the recording was to take place, his wheelchair broke down. With the aid of a technician, Hawking's graduate assistant Chris Burgoyne worked a 36-hour shift to mend the broken wheelchair to ensure Hawking got there, albeit a few minutes late.

He was happy with the episode's good-natured jokes at his expense, although he refused to be portrayed as drunk in the episode's last scene, in which he is discussing astronomy with Homer in Moe's Tavern. "I do remember him changing the scene," Jean recalled. As ever, owing to the idiosyncrasies of his voice synthesizer, some words proved problematic, notably "frutopia", which expressed how, for most Springfield citizens, the hapless efforts of the Mensa elite fell far short of delivering utopia. "I don't know what is the bigger disappointment," Hawking laments of Springfield's leaders: "My failure to formulate a unified field theory – or you."

The episode's final scene sees Hawking and Homer in Moe's Tavern discussing the cosmos, and contains a reference to the universe being toroidal – shaped like a doughnut. (Strangely enough, a 3D doughnut universe has been considered to explain some observations, such as details of the echo of the Big Bang, though it is very much an outsider idea.) The executive producer, Mike Scully, called it "a chance to get the world's smartest man and the world's stupidest man in the same place". "Your theory of a doughnut-shaped universe is intriguing," Hawking tells Homer during the scene. "I may have to steal it."

But when Homer tries to land Hawking with the drinks bill (by imitating his voice synthesizer), the boxing glove comes into its own by landing a punch.

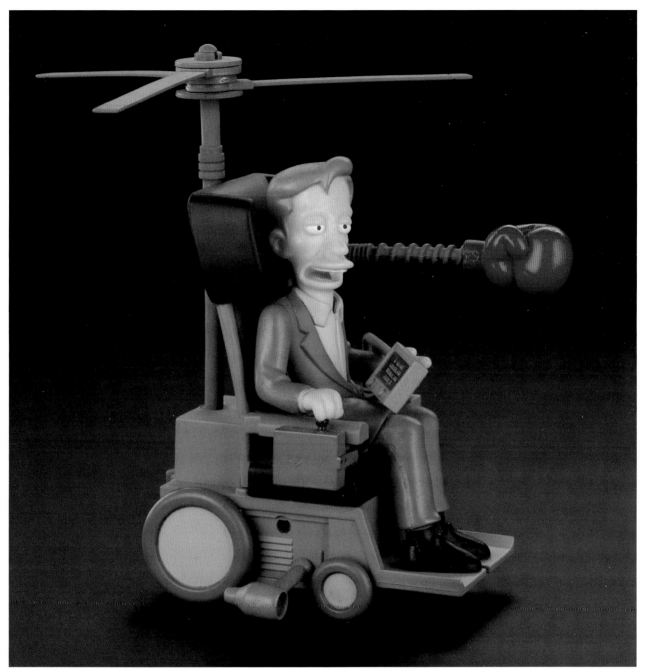

ABOVE Stephen Hawking's "Simpsons' World of Springfield" interactive figure, c. 2003, where his wheelchair includes helpful modifications, from helicopter rotors to a spring-loaded boxing glove.

Hawking's appearances on *The Simpsons* were, collectively, a landmark moment in the representation of disability on screen.

Following "They Saved Lisa's Brain", Hawking guest-starred in more episodes. In 2005, he appeared in season 16's "Don't Fear the Roofer", a parody of the 2001 film *A Beautiful Mind*. In 2007, he featured in the season 18 episode "Stop or My Dog Will Shoot", where he escapes from a maze with his built-in helicopter, and in 2010, during the opening episode of season 22, "Elementary School Musical", he delivers the immortal line, "A brief history of rhyme".

When the Science Museum collected the contents of Stephen Hawking's office, they turned out to include other *Simpsons* memorabilia. There was a baseball jacket from the show, known as a Letterman's jacket to denote a student who has performed well on a varsity team and is given a cloth "letter" to sew on his jacket.

There was also a *Simpsons* clock and a Lisa Simpson Russian doll, the smallest of the family set. The fate of the rest of the family, which can be seen in a photograph taken in 2007, is a mystery, though it is known he would give them to special visitors – in one instance a young boy with cancer – as gifts.

He was routinely invited to *Simpsons*' parties too, attending one with the show's creator Matt Groening, as revealed by this clutch of party passes that hang off a "balloon hook", adorned with the number one, a reference to his first trip in a balloon in 2002.

A SIMPSONS CHARACTER?

In some ways, Hawking's appearances on *The Simpsons* were, collectively, a landmark moment in the representation of disability on screen. By this time, the show was a cultural phenomenon, watched by more than 13 million people in the US alone, who were tuning in to see the world's best-known physicist, who happened to be in a wheelchair, as a guest star.

Though the objects from his office suggest he was a hardened fan, Stephen Hawking was sometimes ambivalent about the publicity attracted by his roles in *The Simpsons*, which may even have overshadowed his other, serious, appearances on film and TV. Asked, during a discussion with the physicist Brian Cox published in a British newspaper, what was the most common misconception he had to endure, Hawking replied, "People think I'm a *Simpsons* character."

SEE ALSO:

To Boldly Go..., p.126
America's Highest Honour, p.168
Will You Be My Valentine?, p.170

OPPOSITE Letterman jacket from *The Simpsons* TV show, from EastWest recording studios, c. 1999–2010. Stephen Hawking's second wife, Elaine, wore the jacket during a hot-air balloon ride to celebrate his sixtieth birthday in 2002. **BELOW LEFT** Framed still from *The Simpsons*' episode "Elementary School Musical". **BELOW RIGHT** VIP passes for the premiere parties of *The Simpsons*' episodes, hanging from a balloon hook that commemorates the balloon flight organized for Hawking's sixtieth birthday.

MABF21 - "Elementary School Musical"
Special Guest Voice by Stephen Hawking

The Universe in a Nutshell

In October 2001, I wrote an article for a national newspaper with the provocative headline: "The answer to the universe and everything?"

The Daily Telegraph was serializing Stephen Hawking's latest book, *The Universe in a Nutshell*, and my story was of course intended to whet readers' appetites with the prospect of a ringside seat for the latest developments in cosmology and physics. The world's best-known physicist thought that scientists "may have" discovered the "theory of everything", referring to what was then known as M-theory, with M standing for "mystery", "membrane" or the "mother of all strings", though Hawking playfully liked to think it could refer to the first name of one of his doctoral students, Marika Taylor, who was working on the subject for him.

A DEEPER THEORY OF EVERYTHING

Since Albert Einstein, the quest for a theory of everything has aimed to blend the theories of the very small (quantum theory) and the very large (relativity) that he helped to pioneer in the first half of the twentieth century, to reveal a deeper theory of everything that would unite all the forces of the universe with a single equation. One of the strangest features of such theories, which Stephen Hawking focused on in his later research, is that they require the universe to have more than three spatial dimensions to unify our picture of all forces and all matter.

However, there have been many false dawns in the quest for a theory of everything. One promising candidate was supersymmetric string theory, in which strings are one-dimensional extended objects, ripples on them are interpreted as particles, and the equations for force and the equations for matter are identical (they are said to have the property of supersymmetry). But, to their surprise, physicists found five possible superstring theories, and were not able to devise experiments to prove which one of them was correct.

Physicists working on this theory subsequently realized that strings were one member of a bigger class of objects called branes. Branes could be extended in more than one dimension, from strings of one dimension to membranes of two dimensions, to those of an arbitrary number – p-dimensions – which were waggishly dubbed p-branes by Professor Paul Townsend, a Cambridge University colleague of Hawking.

Physicists were taken aback in 1995 when at a conference at the University of Southern California Edward Witten pointed out that string theories and p-branes were, in fact, facets of one underlying M-theory, which suggests that we live in a brane world: a four-dimensional surface, or brane, in a higher dimensional mixture of space and time.

M-theory goes way beyond the four dimensions we live in – three of space and one of time – to suggest that there are as many as 11 dimensions. People and most particles inhabit the brane, while the higher dimensions provide a framework to unify all forces, from gravity to those that act between subatomic particles.

Hawking's student at the time, Marika Taylor, worked on "Problems in M-theory" as her thesis, from 1995 to

The answer to the universe and everything?

By ROGER HIGHFIELD
SCIENCE EDITOR

SCIENTISTS may have discovered the "theory of everything", the all-embracing theory of the universe, says Prof Stephen Hawking, the world's best known cosmologist.

"We may have already identified the theory of everything," he claims in his new book, *The Universe in a Nutshell*, referring to what is otherwise known in the scientific world as M theory, with M standing for "mystery", "membrane" or the "mother of all strings".

M theory goes beyond the four dimensions we live in – three of space and one of time – to suggest that there are as many as 11 dimensions. This theory of everything would unite all the forces of the universe with a single equation.

Confirmation of the predictions, encapsulated in the equation's symbols, that this hyperspace is real would mark the greatest achievement of physics and realise Albert Einstein's dream of a single all-encompassing theory of the universe.

Since Einstein, the quest for a theory of everything has depended on combining theories of the very small (quantum theory) and the very large (relativity).

One of the strangest features of such theories is that they require the universe to have more than three spatial dimensions to unify our picture of all forces and all matter.

This is not the first time that scientists have believed that they are on the verge of building a theory of everything.

One promising candidate was supersymmetric string theory, in which strings are one-dimensional extended objects and ripples on them are interpreted as particles. But, to their surprise, physi-cists found five superstring theories and failed to devise experiments to prove which was correct.

After 1985 it became apparent that strings were one member of a bigger class of objects called branes that could be extended in more than one dimension, to membranes of two dimen-
Continued on Page 2

Science: Pages 20 & 21

ABOVE Front page *Daily Telegraph* article on M-theory as a possible "theory of everything", October 2001.

Since Albert Einstein, the quest for a theory of everything has aimed to blend the theories of the very small (quantum theory) and the very large (relativity) that he helped to pioneer in the first half of the twentieth century.

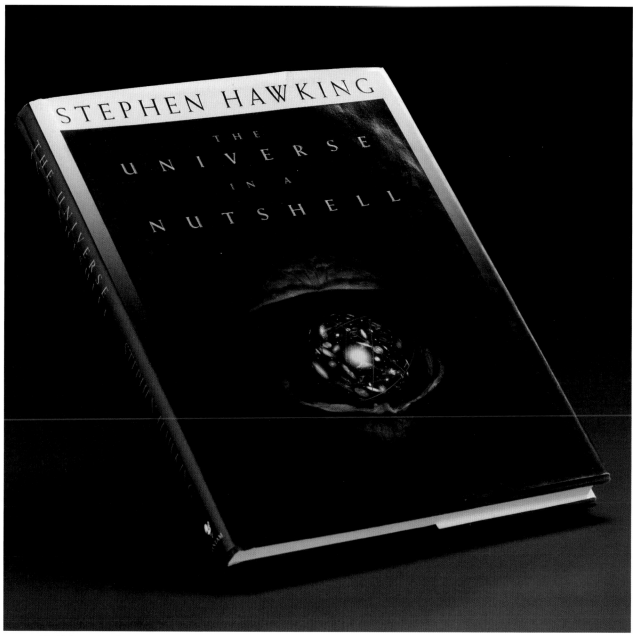

ABOVE *The Universe in a Nutshell*, published in 2001, was Hawking's second major work aimed at a general audience.

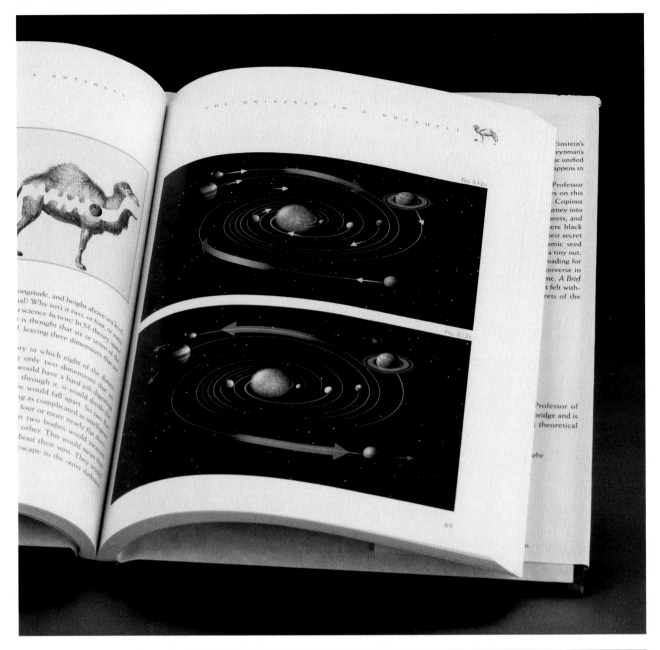

When the book was published in 2001, there was real optimism that M-theory's extra hidden dimensions could be uncovered when the world's largest and most powerful particle accelerator, the Large Hadron Collider (LHC), became operational.

1998. "M-theory was meant to be this overarching theory which explained to you why different types of string theory were related to each other," she said.

The "no-boundary" model of the cosmos that Hawking propounded in *A Brief History of Time* describes the conditions at the birth of the brane world. Though much earlier, it complements M-theory to reveal how the cosmos unfolded. Thus, he contended, M-theory could reveal "the origin and fate of the universe". "Hawking went back and forth on his views on M-theory," recalled Marika Taylor, "but eventually ended up thinking that it may be our best bet for a theory of everything".

BRANE WORLDS

Now, in *The Universe in a Nutshell*, Hawking was discussing how the three dimensions we see could indeed be a brane floating, like a bubble, in a space of half a dozen dimensions. In his interview about the book he told me: "Brane worlds and large extra dimensions could be detected by the next generation of particle accelerators."

When the book was published in 2001, there was real optimism that M-theory's extra hidden dimensions could be uncovered when the world's largest and most powerful particle accelerator, the Large Hadron Collider (LHC), a 27-kilometre underground ring of superconducting magnets near Geneva, became operational. M-theory rests on a concept called supersymmetry, which predicts that every known particle has a supersymmetrical equivalent, and the hope was that the LHC could find these equivalent particles, or superpartners. "Brane worlds and large extra dimensions could be detected," said Hawking.

Confirmation that this hyperspace is real would mark the greatest achievement of physics, and realize Albert Einstein's dream of a single all-encompassing theory of the universe. "This would make quantum gravity an experimental science," added Hawking.

Despite his optimism all those decades ago, M-theory has not aged well, alas. Previous experiments at another particle accelerator, the Large Electron-Positron Collider, which operated from 1989 to 2000 in the same tunnel as the LHC, had already cast doubt on the simplest supersymmetric models. Then scientists had to wait for the LHC, which started operations in September 2008, later than expected. As data from the LHC has continued to accumulate, models of supersymmetry that were once thought plausible by the physics community have been gradually undermined by the lack of experimental data to support them.

The absence of evidence for supersymmetry at the LHC does not, however, mark its death knell. Some physicists maintain that the most plausible theories predict the existence of superpartners that are too heavy to be

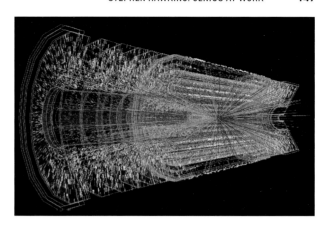

discovered with the current generation of particle accelerators. Others believe that superpartners are created in collisions at the LHC, but that they escape detection because they do not manifest in the way scientists predicted. Most searches for these new particles assume that they would decay soon after being created, but there are alternative theories that predict long-lived superpartners which could travel between a few micrometres or even 100,000 kilometres before eventually decaying. "People still work on this," said Marika Taylor, "but they just don't call it M-theory".

By the end of this decade, an upgrade to the LHC known as the High-Luminosity LHC will allow experimentalists to explore more uncharted areas of the supersymmetric landscape to see if the optimists are right, or if we need to go back to the drawing board. Ultimately, all the answers lie in nature, and the only way to test supersymmetry is to keep looking.

Back in 2001, I reminded Stephen Hawking that he had been accused of hubris when he wrote in *A Brief History of Time* that we would one day "know the mind of God" and had boldly given the odds as 50/50 that a theory of everything would emerge by the millennium. Did he want to give new odds?

"I thought we might find the theory of everything by the end of the twentieth century," he replied. "However, although we made a lot of progress, our goal still seems about the same distance away. I have revised my expectations downwards, but I still think there's a good chance of finding it by the end of the century, only now it is the twenty-first century."

OPPOSITE A page from *The Universe in a Nutshell* illustrating the consequences of space with more than three dimensions. M-theory suggests there are as many as nine or ten dimensions, but that most of these are "curled up" and so are not apparent to us. **ABOVE** Simulation of collision between two protons in the LHC's ATLAS Experiment.

Universe on a Beach Ball

Tucked behind the computer monitor on Stephen Hawking's desk sat this incongruous and inflatable intruder.

The ball, which measures just under a foot across and looks like a gaudy toy, is a piece of merchandise sold by the American space agency NASA. For Stephen it was something he often gave as a present to young visitors to his office, including his grandchildren.

But this throwaway curiosity also had immense significance to professors, because it celebrated one of the marvels of the modern scientific age: a feat of observation that ended a decades-long argument about the nature of the universe, and confirmed that our cosmos is much stranger than we ever imagined. More to the point, it harked back to a milestone meeting Stephen Hawking had co-organized in Cambridge two decades earlier, at which similar insights into how structure emerges in the universe had been thrashed out.

THE AFTERGLOW OF THE BIG BANG

This gaudy globe is decorated with a picture of the baby universe. Seeing it bounce around at family gatherings must have given Hawking huge satisfaction at how this picture had vindicated his ideas. Predictions are ten a penny in cosmology, and the field is only aware of real advances when they have convincing experimental evidence. That evidence had come in the form of an image of the afterglow of the Big Bang that adorns this ball.

Formally called the cosmic microwave background, it was captured in an image taken in 2003 by NASA's Wilkinson Microwave Anisotropy Probe (WMAP), a satellite named in honour of the cosmologist David Wilkinson of Princeton University. There are certain places, called Lagrange points, to which objects like this probe or the more recent James Webb Space Telescope are sent because they tend to stay put there: the combined forces acting on them – gravitational and centrifugal – balance out to create a kind of celestial parking spot.

The results from WMAP, now printed onto the beach ball, show "when cosmology became a precision science", as Stephen Hawking put it. While the outline of the map

OPPOSITE Britt Griswold, who works on the Wilkinson Microwave Anisotropy Probe (WMAP), put WMAP's famous image of the early microwave universe, the afterglow of the Big Bang, on a beach ball.

had been glimpsed before, in April 1992 by NASA's COBE satellite, it was put on a more quantitative footing by the WMAP satellite a decade later, and subsequently by the European Space Agency's Planck satellite in 2013.

It was the leader of the WMAP mission, Chuck Bennett, who had the playful idea of putting its famous cosmic image on a beach ball, and his colleague Britt Griswold who made it happen. Aside from in Stephen Hawking's office, the ball made various high-profile appearances: at his seventieth birthday exhibition in the Science Museum, for example, and in Leonard and Sheldon's apartment in *The Big Bang Theory*, the hit TV comedy show.

THE FURTHEST WE CAN SEE

The picture on the ball's surface is nothing less than a snapshot of the universe at about three hundred-thousandths of its present age. It represents the furthest we can see in microwave light, the first and therefore oldest visible light in the universe, created just 378,000 years after the Big Bang, which took place around 13.8 billion years ago before planets, stars or galaxies existed. As the cosmos expanded, this light was stretched by the very fabric of space-time, so it is now longer-wavelength microwave light, invisible to our eyes but not to the sensitive instruments of WMAP.

The colours on the ball indicate the temperature variations of light within the young universe, with red standing for hotter and blue for cooler. The extraordinarily evenly dispersed microwave light that now bathes the universe averages a very frigid 2.73 degrees above absolute zero, which is minus 273°C, the lowest possible temperature. WMAP was sensitive enough to resolve these tiny temperature fluctuations, which vary by only millionths of a degree. The sizes of the hot and cold spots enabled scientists to calculate fundamental values for the shape, size, age and rate of expansion of our universe, along with its contents. The prominent red stripe visible around the equator is a more recent, closer and hotter foreground signal that comes from our own Milky Way galaxy.

DARK MATTER, DARK ENERGY

This inflatable ball, Stephen Hawking's computer, the air we breathe, the distant stars – all are made up of atoms that are constellations of subatomic particles: protons, neutrons and electrons. Astronomers like to call all

Table 6
WMAP Seven-year CMB Dipole Parameters

$d_x{}^a$ (mK)	d_y (mK)	d_z (mK)	d^b (mK)	l	b
-0.233 ± 0.005	-2.222 ± 0.004	2.504 ± 0.003	3.355 ± 0.008	263.99 ± 0.14	48.26 ± 0.03

Notes. The measured values of the CMB dipole signal. These values are unchanged from the five-year values and are reproduced here for completeness.
[a] Cartesian components are give in Galactic coordinates. The listed uncertainties include the effects of noise, masking and residual foreground contamination. The 0.2% absolute calibration uncertainty should be added to these values in quadrature.
[b] The spherical components of the CMB dipole are given in Galactic coordinates and already include all uncertainty estimates, including the 0.2% absolute calibration uncertainty.

Values of $v_{\rm eff}$, $\Omega_{\rm eff}$, and Γ for a point source with $\beta = -2.1$, typical for the sources in the WMAP point source catalog, are given in Table 2.

4. SKY MAP DATA AND ANALYSIS

The seven-year sky maps are consistent with the five-year maps apart from small effects related to the new processing methods. Figure 5 displays the seven-year band average Stokes I maps and the differences between these maps and the published five-year maps. The difference maps have been adjusted to compensate for the slightly different gain calibrations and dipole signals used in the different analysis. The small Galactic plane features in the K-, Ka-, and Q-band difference maps arise from the slightly different calibrations and small changes in the effective beam shapes. Pixels in the Galactic plane region observed over a slightly smaller range of azimuthal beam orientation in the current data processing relative to previous analyses, resulting in slightly less azimuthal averaging and thus a slightly altered effective beam shape. The calibration changes relative to the five-year data release are small, all $<0.5\%$ as indicated in Table 1.

Temperature and Polarization Power Spectra

The Temperature Dipole and Quadrupole

Our CMB dipole value was obtained using a Gibbs method (Hinshaw et al. 2009) to estimate the dipole from the five-year ILC map and foreground reduced uncertainties on the measured parameters were set to results as an estimate of the effect of residual. The dipole measured from the seven-year significant changes from that obtained from to the best-fit dipole parameters remain five-year values, and are presented in

likelihood value of magnitude of the $+1)C_l/2\pi = 197^{+2972}_{-155}\,\mu K^2(95\%\ CL)$ on the analysis of the ILC map using is essentially unchanged from the below the most likely value predicted. This value, however, is not particcontribution of values predicted by the et al. (2011).

n-year band average maps for for all five WMAP frequency on is evident in all frequency ar scale features visible in gree in the V-band maps, are poorly constrained by the dominate the appearance

of the map, they are properly de-weighted when these maps are analyzed using their corresponding Σ^{-1} matrices, so useful polarization power spectra may be obtained from these maps. The relatively large amplitudes of these modes limits the utility of using difference plots between the five-year and seven-year map sets to test for consistency.

Figure 5. Plots of the Stokes I maps in Galactic coordinates. The left column displays the seven-year average maps, all of which have a common dipole signal removed. The right column displays the difference between the seven-year average maps and the previously published five-year average maps, adjusted to take into account the slightly different dipoles subtracted in the seven-year and five-year analyses and the slightly differing calibrations. All maps have been smoothed with a 1° FWHM Gaussian kernel. The small Galactic plane signal in the difference maps arises from the difference in calibration (0.1%) and beam symmetrization between the five-year and seven-year processing. Note that the temperature scale has been expanded by a factor of 20 for the difference maps.

Figure 6. Plots of the seven-year average Stokes Q and U maps in Galactic coordinates. All maps have been smoothed with a 2° FWHM Gaussian.

4.1.2. Low-ℓ W-band Polarization Spectra

Previous analyses (Hinshaw et al. 2009; Page et al. 2007) exhibited unexplained artifacts in the low-ℓ W-band polarization power spectra demonstrating an incomplete understanding of the signal and/or noise properties of these spectra. Specifically, the value of C_7^{EE} for $\ell = 7$ measured in the W band was found to be significantly higher than could be accommodated by the best-fit power spectra, given the measurement uncertainty. This result, and several other anomalies, caused these data to be excluded from cosmological analyses. Significant effort has been expended trying to understand these spectra with the goal of eventually allowing their use in cosmological analyses.

A set of null spectra was formed based on the latest uncleaned W-band polarization sky maps to test for year-to-year and DA-to-DA consistency. Polarization cross power spectra were calculated for pairs of maps using the Master algorithm (Hivon et al. 2002) utilizing the full Σ^{-1} covariance matrix to weight the input maps. The polarization analysis mask was applied by marginalizing the Σ^{-1} over pixels excluded by the mask to minimize foreground contamination. Appropriately weighted null signal combinations of these spectra were formed to determine if any individual years or DAs possessed peculiar characteristics. The uncertainties on the power spectra were evaluated using the Fisher matrix technique (Page et al. 2007) and measured map noise levels. Since the input sky maps contain both signal (mostly of Galactic origin) and noise, an additional

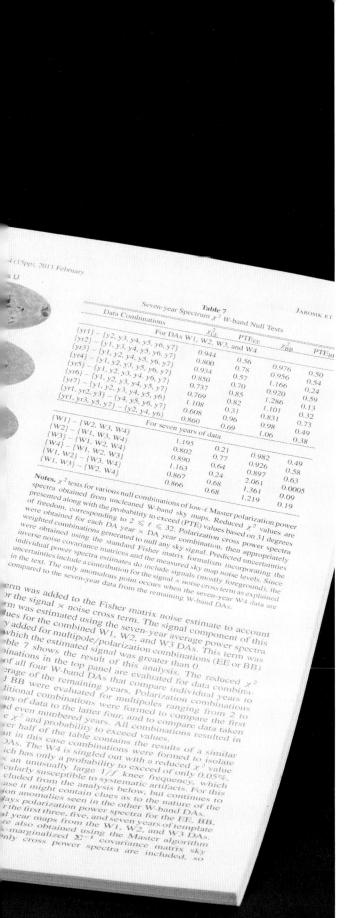

material constituted in this way "baryonic matter". Once upon a time, they thought almost the entirety of the universe was composed of this baryonic matter, but in the past few decades, ever more evidence has accumulated to suggest that the majority of the universe consists of something else, though exactly what remains somewhat mysterious. WMAP revealed that only a measly 5 per cent of the universe consists of atoms, with the rest consisting of stuff that has never been directly detected in the laboratory. Twenty-five per cent is what is called "cold dark matter", an unidentified source of gravity, and 70 per cent "dark energy", which has a gravitationally repulsive effect that explains the "flatness" of the universe. This is a way of expressing how two light beams shooting side by side through space will stay parallel forever.

Before the discovery of this dark energy, scientists expected that the gravitational pull of all matter in the universe would eventually make it collapse into a "big crunch". However, observations since the 1990s show this is not the case. Because all the matter and energy in the universe, including dark matter and dark energy, adds up to the concentration at which the energy of the outward expansion balances the energy of the inward gravitational pull, space extends flatly in all directions. Our universe is indeed very nearly flat and must therefore come very close to this critical density, which is calculated to be around 5.7 hydrogen atoms' worth of stuff per cubic metre of space.

Like cosmic detectives, the WMAP team compared the unique "fingerprint" of patterns imprinted on this ancient light with fingerprints predicted by various cosmic theories and found a match. The dappled irregularities depicted on the beach ball were predicted by "inflationary theory", a concept postulated in 1980 by Alan Guth, then at Stanford University. It was at the seminal workshop hosted by Stephen Hawking in Cambridge in 1982 that the details of how these irregularities arose were worked out.

Inflationary space and time started with the Big Bang, the early universe exploding into being faster than the speed of light from a size smaller than that of a subatomic particle. From a fraction of a second after the Big Bang this mind-boggling growth spurt flattened out the visible universe like a vast sheet snapped tight, leaving everything looking relatively uniform in all directions. Cosmic inflation also yields pristine flatness.

The tiny irregularities in the microwave background shown on the ball mean, however, that some regions of the universe had slightly higher density than others, along

LEFT The *Astrophysical Journal* issue showing the results of seven years of observations of the cosmic microwave background by the WMAP probe, 2011.

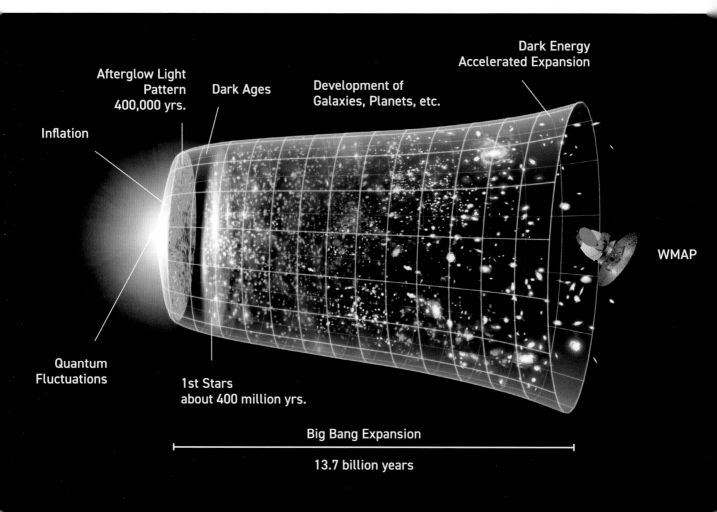

Inflation

Afterglow Light Pattern 400,000 yrs.

Dark Ages

Development of Galaxies, Planets, etc.

Dark Energy Accelerated Expansion

WMAP

Quantum Fluctuations

1st Stars about 400 million yrs.

Big Bang Expansion

13.7 billion years

ABOVE The universe's evolution from the Big Bang (left) to the present day (right), when the afterglow of the Big Bang was observed by the WMAP probe. As the universe cooled immediately after the Big Bang, atomic nuclei began to form and combine with electrons to form atoms. Within a million years dense regions of atoms attracted nearby matter, forming gas clouds and, after about 400 million years, stars and then small galaxies. These small galaxies merged into larger galaxies and eventually became the Solar System as we know it.

with slightly higher gravitational attraction. This slowed the expansion of those regions, and could eventually cause them to collapse and coalesce into galaxies and stars. Stephen himself would remark that WMAP's evidence for inflation was at that time the most exciting development ever to have taken place in physics.

ARE THOSE HIS INITIALS?
According to Michael Turner of the University of Chicago, one of those present at that 1982 workshop, there was another reason for Stephen in particular to like the WMAP image of the infant universe. Among the splodges of

colour, he tweeted, "you can see his initials in blue." Turner is referring to the way in which the swirl of blue and green depicting the microwave background appears to coalesce in one place into "SH", Stephen Hawking's initials. Turner went on to email me: "Many have commented about this :-) Great product placement. His name is written across the sky forever!" Seriously, though, if something as apparently unlikely as Hawking's initials can be found in these data, then the chances of finding other apparently improbable patterns may also result from pareidolia, the all-too-human tendency to perceive a meaningful image in a random or ambiguous pattern.

In the past few decades, ever more evidence has accumulated that suggests the majority of the universe consists of something that remains somewhat mysterious.

Hawking's old collaborator, Sir Roger Penrose, for example, believes there is evidence that the microwave background contains anomalies he has dubbed "Hawking Points", that are evidence to support his own idea of "conformal cyclic cosmology". This proposes that our universe is just one in an infinite series. Although it is based on an analysis with colleagues published in the *Monthly Notices of the Royal Astronomical Society*, rather than just the evidence of his own eyes, many of his peers remain unconvinced.

But the link between small-scale quantum fluctuations and the overall look of the universe today has held true. "Over several decades, Stephen had proposed how quantum fluctuations in the early universe could lead to large-scale structure or anisotropies in the cosmic microwave background radiation," said Juan-Andres Leon, the Science Museum's curator of Stephen's office collection.

Decade after decade he was awaiting observational proof that the universe had these anisotropies, and in general the WMAP data fits. This is one of the predictions made by Stephen Hawking and a generation of scientists about what the early universe might look like.

"Some people say he was more excited about this than he was about black holes," added Juan-Andres. "If he had lived to get a Nobel Prize, it might well have been for this work." No wonder, then, that among the handful of journals in his office there is a 2011 copy of the *Astrophysical Journal*, which summarizes seven years of WMAP observations.

The beach ball shows a different side to Hawking's science. He is usually fêted for his revolutionary discovery that black holes give off radiation, now called Hawking radiation in his honour. But both preoccupations highlight his creative trademark as a theoretician: his pursuit of the unification of quantum mechanics with general relativity, where the latter is the theory of the very big and gravity, and the former the theory of the very small, and the world of subatomic particles and atoms. Hawking's Cambridge

colleague Paul Shellard assesses the difference between them:

The distinction that can be drawn between Stephen's fundamental advances on quantum black hole radiation and that on the quantum origin of structure in the universe (independently of others, some in the Soviet Union), is that the former is unlikely to be tested experimentally soon, while the latter has strong observational evidence – that is, is vindicated already.

FINGERPRINTS OF CREATION

The little inflatable ball is not the only celebration of this milestone preserved among Stephen Hawking's effects. There is also a small glass apple – a gift from researchers at Intel and a nod to Hawking's position as Lucasian Professor at Cambridge, a post formerly held by Sir Isaac Newton – that has been painted to evoke the cosmic microwave background, which contains what Hawking called the "fingerprints of creation". Both tell a remarkable story about the power of human imagination, in this case his predictions. "Look carefully at the map of the microwave sky and it is the blueprint for all the structure in the universe," he once said. "We are the product of quantum fluctuations in the very early universe. God really does play dice."

SEE ALSO:

Hawking Radiation, p.46
Workshop on the Baby Cosmos, p.58
The Anthropic Principle, p.100
Beyond the No Boundary, p.156
Hawking in Space, p.164
Stephen Hawking, You Are Wrong!, p.176
Hawking Sings Monty Python, p.194

Beyond the No Boundary

Stephen Hawking was a supporter of the African Institute for Mathematical Sciences, AIMS, a pan-African network of centres of excellence founded in 2003 by his friend Neil Turok, a South African physicist. This buffalo horn was Turok's modest way of saying thank you. "I just brought him this little souvenir as a present when I came back from teaching at AIMS in Senegal one year," said Turok, who is now at the University of Edinburgh. "It's an embarrassingly small gift – I owed him much more." In 2008 Hawking visited AIMS's South African centre, there meeting Nelson Mandela, the anti-apartheid activist and politician who served as the first president of South Africa.

Back in the UK, Hawking's Cambridge office was for Turok a special place where "we had wonderful times together." That does not mean, however, that they always saw eye to eye, even if in physics, unlike some fields of science, disagreements tend to be amiable.

CYCLIC UNIVERSE
In 2002, Turok had proposed a view of the cosmos where time has no beginning or end. Instead, he argued, the cosmos undergoes cycles of expansion and contraction so that it endlessly dies and rises from its ashes. In this cyclic universe theory, there is no cosmic inflation – a faster-than-light expansion of the universe that spawned many others – and no accompanying gravitational waves.

Hawking, who had supported and expanded on the theory of inflation since the 1980s, disagreed with Turok's theory, and had even publicly bet Turok that the Planck satellite that maps the microwave "echo" of the Big Bang would detect primordial gravitational waves. "It didn't," said Turok.

More fundamentally, in 2017 Turok attacked the "no-boundary proposal" Hawking and Jim Hartle had come up with, in which they envisaged the so-called "wave function of the universe" that encompasses the entire past, present and future at once. By the terms of this proposal, asking what came before the Big Bang would be, as Hawking famously remarked, "like asking what lies south of the South Pole".

"Although we eventually showed the no-boundary proposal didn't work," said Turok, "he continued to invite me to his private 'retreats' and to help me build AIMS."

STEPHEN'S INFLUENCE LIVES ON!
Since then, even though most cosmologists still support inflation, Turok has come to believe that the no-boundary idea was more visionary than even Hawking had realized. "The problem was Stephen tried to implement it within inflationary cosmology, where it didn't work. For the past couple of years, we have been implementing what is in effect his idea, in a much more minimal and powerfully predictive framework. Stephen's influence lives on!" As ever, the final judgement on inflation will come from new observations of the heavens.

SEE ALSO:

Hawking Makes Waves, p.40
Universe on a Beach Ball, p.150
A World of Souvenirs, p.200

OPPOSITE Carved buffalo horn gifted to Hawking by Neil Turok.

In 2008 Hawking had even visited AIMS's South African centre, there meeting Nelson Mandela, the anti-apartheid activist and politician who served as the first president of South Africa.

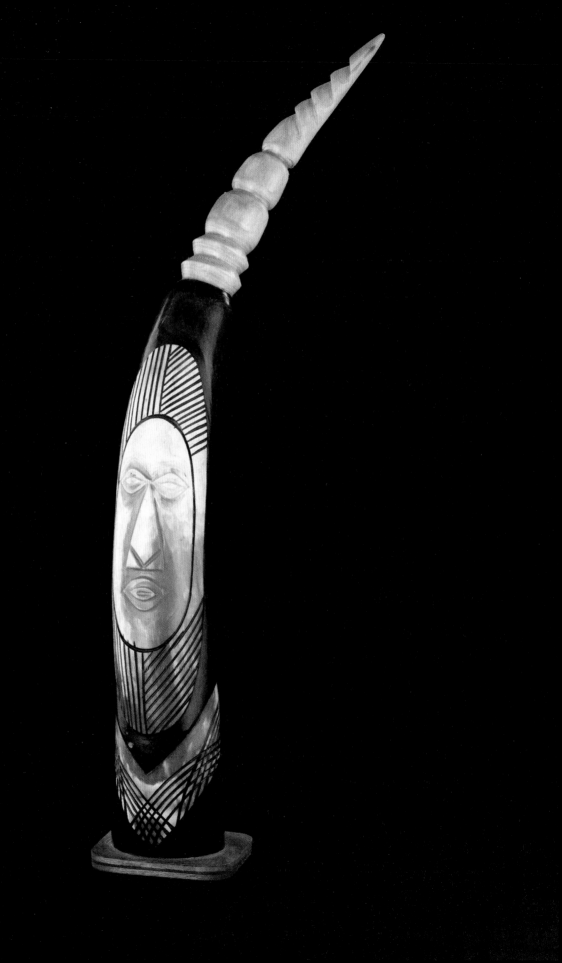

Groovy Icon, Serious Science

There are some objects in the office which have a subtle scientific significance that can easily be overlooked. They include this lava lamp, which could often be spotted in the background of photographs taken during media interviews. "The lava lamp was given to my father for Christmas in the mid-noughties by my brother Tim Hawking," said his daughter, Lucy. "Tim thought it was a fun object which gave an otherworldly glow to the office and made one think of extra-planetary landscapes."

A symbol of Hawking's playful 1960s/1970s generation, the decorative lava lamp was invented in 1963 by the British entrepreneur Edward Craven Walker, the founder of the lighting company Mathmos in Poole, Dorset. It contains two coloured liquids that are insoluble and of similar density. The heavier liquid absorbs heat from a light bulb and, as it expands, it becomes less dense, and rises. But as the liquid rises, it gradually cools, becoming denser and therefore heavier, so it eventually sinks. And so the cycle begins once more.

PLANETARY PROCESSES
Similar processes occur within planets. For example, a recent computer model suggested that sand and mud subducted off the coast of California around 75 million years ago returned to the Earth's crust by rising up through the mantle like blobs in the lamp. Pluto's icy surface is renewed by the same process of convection, replacing older ices with fresher material, according to an analysis of data from the New Horizons space probe.

"Lava lamps also feature in the *George* book series by Stephen and Lucy Hawking, where they are used to describe geological processes," added Juan-Andres Leon, the Science Museum's physics and Hawking curator.

SEE ALSO:
—————
Stephen's Secret Key to Childhood Wonder, p.160
To Pluto, and Beyond, p.186

OPPOSITE Green lava lamp, made by Mathmos in Poole, Dorset.

"The lava lamp was given to my father for Christmas in the mid-noughties by my brother Tim Hawking," said his daughter, Lucy. "Tim thought it was a fun object which gave an otherworldly glow to the office and made one think of extra-planetary landscapes."

Stephen's Secret Key to Childhood Wonder

Not content with taking the world of non-fiction publishing by storm with *A Brief History of Time*, Stephen Hawking co-authored a series of children's books. Inspired by his grandson, William, they were a collaboration with his daughter, Lucy. "This is, I think, the first time you have a world-famous expert in science writing non-fiction for such a young age group, using a dramatic storyline," said Lucy. "According to Penguin, our publisher, our very first book, *George's Secret Key to the Universe*, helped to kick-start non-fiction writing by experts for this age range."

The idea of documenting a boy's adventures in the universe came to Lucy while watching her father at William's ninth birthday party, where his young friends were held rapt by his science. "Suddenly everything was abandoned, from the cake to the bouncy castle," she recalled. "They were talking to him about black holes and the end of the universe, and what would happen if they flew too close to the sun, or if the sun were to disappear from the sky." Her father was "really good at giving clear answers that they could understand", Lucy said. "One little boy asked Dad what would happen if he fell into a black hole, and he replied, 'You'd be turned into spaghetti.'"

Afterwards, Lucy searched for a book for her son and his friends that would explain cosmic concepts in a scientifically rigorous way, and yet used intriguing concepts like spaghettification. She "found nothing".

The black hole question, in particular, put William's little friend at the heart of the story, and it was this that planted the seed of an idea: for her and Stephen to write books together that "could answer the questions that kids were asking in a storytelling format". Lucy was an experienced writer and her arts background made her a perfect foil for her father. He had never written a dramatic storyline about astrophysics, and this would require him to frame his thoughts in a different way.

They began the first draft during a summer visit to Hong Kong, and when they aired the idea of him co-authoring a kids' book at a press conference, Lucy said there was an eruption of interest – "the whole event went nuts" – and they knew they were on to something: "Bada-boom!"

HAWKING'S ALTER EGO

The narrative of the book is driven by a young boy, George Greenby, who on meeting his new neighbours Eric and Annie begins an adventure through the universe. Eric owns the world's most intelligent supercomputer, which is the portal that allows George to visit the edge of the Solar System and beyond. Stephen and Lucy named the supercomputer Cosmos, after the real COSMOS supercomputer in the Centre for Theoretical Cosmology at the University of Cambridge, much to the delight of his former student Paul Shellard, who coordinates the COSMOS effort.

In the character of Eric, a brilliant scientist, there is a poignant glimpse of Hawking himself as a young man,

OPPOSITE The first in the children's book series authored by Stephen and his daughter Lucy Hawking, *George's Secret Key to the Universe*, 2007.

The idea of documenting a boy's adventures in the universe came to Lucy while watching her father at William's ninth birthday party, where his young friends were held rapt by his science.

BLACK HOLES • LIGHT & STARS • MASS • EARTH • COMETS
ASTEROID BELT • THE SOLAR SYSTEM • MATTER • NEUTRON STAR • TEMPERATURE •
THE MOON • BLACK HOLES • THE PLANETS • THE NIGHT SKY •
LIGHT & STARS • MASS • EARTH • COMETS • THE SOLAR SYSTEM • MATTER •
Includes the LATEST IDEAS about BLACK HOLES!
GEORGE'S SECRET KEY TO THE UNIVERSE
LUCY & STEPHEN
HAWKING

In the character of Eric, a brilliant scientist, there is a poignant glimpse of Hawking himself as a young man, before the onset of his disease.

before the onset of his disease. As Lucy explained when the book was published, "He [Hawking] loves the idea of an alter ego who actually gets to visit outer space rather than just thinking about it all the time."

The universe George experiences during his travels is an accurate portrayal of the reality, supported by the latest data gathered by astronomers and cosmologists. In recognition of the birthday party that inspired them, Stephen and Lucy wanted their young audience to learn something tangible about the world of science, and so dealt with questions such as what it would be like to be on a planet with a huge gravitational tug.

George's Secret Key to The Universe was the first fruit of the father–daughter writing endeavour, aided by Stephen's French graduate student Christophe Galfard and illustrator Garry Parsons. Their protagonist's first cosmic adventure combined an action-packed story with child-friendly essays by Hawking himself and other researchers.

SUCH A GUFFAW

Writing turned out to be a bonding experience, with father and daughter meeting frequently over dinner at his Cambridge home, as ever in the company of carers. One time she read him a passage she had just written for their second book, *George's Cosmic Treasure Hunt*, in which George's grandmother appears.

Lucy had modelled the character on her father's mother, Isobel, and, on realizing the resemblance during a dramatic entrance – peppered with loud protestations that everyone else was wrong and absolutely everything had to be done her way – Stephen "gave such a guffaw of laughter", recalled Lucy, that he looked as if he would fall out of his chair. "I have never seen that happen before. Two carers had to jump in front of him and grab him. That's when I knew he was hooked; that he'd started to enjoy this."

Their writing partnership would evolve. While for the first book Hawking saw himself as a consultant, ensuring it was scientifically accurate, by the third he even came up with the plot, in which a mysterious group want to destroy the Large Hadron Collider – the atom smasher based just outside Geneva which can recreate conditions as they were just after the Big Bang of creation – and thereby stop the secrets of the early universe becoming known.

Hawking came up with this plot, said Lucy, because he was "intrigued and amused" by the rumours swirling around the start-up of the Large Hadron Collider in 2008 – not least that it would spawn a planet-eating black hole. "By that time he was into the adventure, and by then was saying things like, 'I don't think George would say that.'"

GEORGE'S ADVENTURES

Over the years, the original idea for just one book developed into a series of adventures for George, taking him across the universe. Over a further five books, George and Annie journey through space and time on quests that bring the very real wonders of the universe into the lives of young readers. The last, *George and the Ship of Time*, is perhaps the gloomiest, according to Lucy, since it was inspired by her father's increasing unease about the future of humankind, whether you considered the impact of AI, for example, social inequality or climate change.

However, for this book Stephen Hawking was not the co-author. "He was still alive while I was writing it," explained Lucy, "but he was too unwell, and too focused on trying to use the time remaining to him on his high-level scientific projects. For this reason, I asked for his name not to feature on the front cover, as I felt it would be misleading." But thanks to George, over a dozen years Stephen Hawking wrote more books with his daughter Lucy than he did with anyone else, all of which have been translated into dozens of languages, from Zulu and Xhosa to Mongolian.

After his death in 2018, Lucy visited his house, and found herself moved to tears by the sight of a tablecloth she had bought for him in New Delhi while she was on a tour of South Asia to promote one of the George books. "As my father liked to say, 'Where Harry Potter has magic, we have science.'"

SEE ALSO:

Groovy Icon, Serious Science, p.158
Time Travellers' Party, p.172

OPPOSITE Hawking with his daughter and collaborator Lucy in his office.

Unlocking the Universe

The six George books had featured short, bright essays by well-known scientists, from the AI guru Demis Hassabis to volcanologist Tamsin Mather. "None of them had written for this audience before," recalled Lucy. After her father's death, Lucy had the idea of bringing them all together in *Unlocking the Universe*, a non-fiction anthology for young adults that featured several of Hawking's long-term friends, including Kip Thorne, who contributed an essay on "Wormholes and Time Travel", and Lord Rees, Astronomer Royal, who asked, "Is Anyone Out There?" In his own two-page essay that opens the collection, Hawking himself talked of the creation of the universe, and how we owed our very existence to variations in the baby cosmos: "If the early Universe had been completely smooth, there would be no galaxies or stars, and so life couldn't have developed."

ABOVE Tamsin Mather, Oxford volcanologist and contributor to *George and the Blue Moon* (2016).

Hawking in Space

Stephen Hawking wanted to reach for the stars. On his sixtieth birthday, he took a trip in a hot-air balloon that had been adapted for his wheelchair and spent an hour or so floating across Cambridgeshire. In 2007, on his sixty-fifth birthday, he told me, "This year I'm planning a zero-gravity flight, and to go into space in 2009." Concerned that the future of the human race would depend on exploring the cosmos, he wanted to show humanity the way to the heavens.

No wonder, then, that his office contained all sorts of space-related paraphernalia, from a Dorling Kindersley volume on space flight ("The complete story from Sputnik to Shuttle – and beyond") to books by Buzz Aldrin, along with an Apollo XI poster celebrating Aldrin and Neil Armstrong becoming the first to walk on the Moon in 1969. There was also a mug commemorating Principia, the six-month mission by European Space Agency astronaut Tim Peake on the International Space Station in 2015–16

(the Soyuz spacecraft that brought him back to Earth is on display in the Science Museum).

A FLIGHT IN A VOMIT COMET

Taking pride of place on a wall of Hawking's office near his photograph with Pope Francis was a photograph taken during his flight in a "Vomit Comet", a present from the then NASA administrator Charlie Bolden, framed along with some flags that had been taken into space, and a NASA aerial photograph of Cambridge.

A "Vomit Comet" is the name astronauts have come to give to a zero gravity, "parabolic" flight in which an aeroplane uses a roller-coaster-style flight path to create the condition of weightlessness. Initially, the plane is pulled up to approximately 45 degrees (nose high) and, as it goes over the top of its parabolic trajectory for the next 25–30 seconds, everything in the plane becomes weightless. This weightlessness is akin to that experienced during free fall when sky diving.

At approximately 30 degrees (when the nose is down), a gentle pull-out is started, and the g-force rises smoothly to about 1.8 times the force of gravity until the aircraft reaches a flight altitude of 24,000 feet where, at the top of the arc, you can begin to free-fall again.

Stephen Hawking's flight was organized in 2007 by Peter Diamandis, CEO of the Zero Gravity Corporation, in a specially modified Boeing 727-200 aircraft named G-FORCE ONE. "It was one of the proudest moments of my life," recalled Diamandis. But it did not prove straightforward. The US Federal Aviation Authority declared Hawking unfit to fly. An irritated Diamandis therefore asked several doctors for signed letters to submit to the FAA stating, "without question, that Hawking was able-bodied for a zero-g flight", and paid for malpractice insurance to cover them. For extra reassurance, Diamandis set up an emergency room on board G-FORCE ONE with four doctors and two nurses. He planned for just one or two 30-second parabolas, to give Hawking the briefest sensations of weightlessness.

"The first parabola went so smoothly, and Hawking was having such a great time, that we flew a second, and a third … and another and another," recalled Diamandis. "In total, we made eight arcs with him aboard." Making acrobatic low-g flips during these arcs like a "gold-medal gymnast", as one crew member put it, Stephen Hawking took blissful leave of his wheelchair. "It was amazing," he said. "I could have gone on and on – space, here I come!" In 2014, in a

message to NASA astronauts on the International Space Station, Hawking fondly recalled his zero-g experience. "People who know me well say that my smile was the biggest they'd ever seen."

SPACE AND SURVIVAL

But there was a serious side to his fascination with space flight. "Life on Earth is at ever increasing risk of being wiped out by a disaster," he said, "such as sudden global warming, nuclear war, a genetically engineered virus or other dangers. I think the human race has no future if it doesn't go into space."

The founder of Virgin Galactic, Sir Richard Branson, wanted to help Hawking achieve that dream, and in 2008 offered him a free ride on Virgin's suborbital SpaceShipTwo space plane. At that time, Branson expected to use his SpaceShipTwo to carry six passengers at a time up to altitudes of 110,000 metres, comfortably above the altitude where orbital dynamic forces become more important than

ABOVE Poster commemorating the fortieth anniversary of the first moon landing, discoloured from being next to the window in Hawking's office.

OPPOSITE Hawking in NASA's "vomit comet", the zero-gravity flight that follows a parabolic trajectory so that everything inside the aeroplane becomes weightless.

ABOVE NASA Administrator Charles Bolden presents a commemorative NASA montage to Professor Stephen Hawking in his office, 2015.

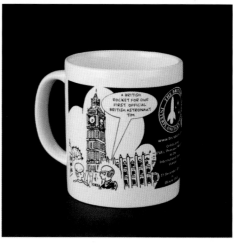

ABOVE LEFT Poster of Virgin Space Ship *Unity*, featuring Hawking's iris. **ABOVE RIGHT** Mug made in celebration of the Principia mission, the name given to European Space Agency astronaut Tim Peake's six-month mission stay on the International Space Station, launched in December 2015 to maintain the weightless research laboratory and run scientific experiments for hundreds of researchers on Earth.

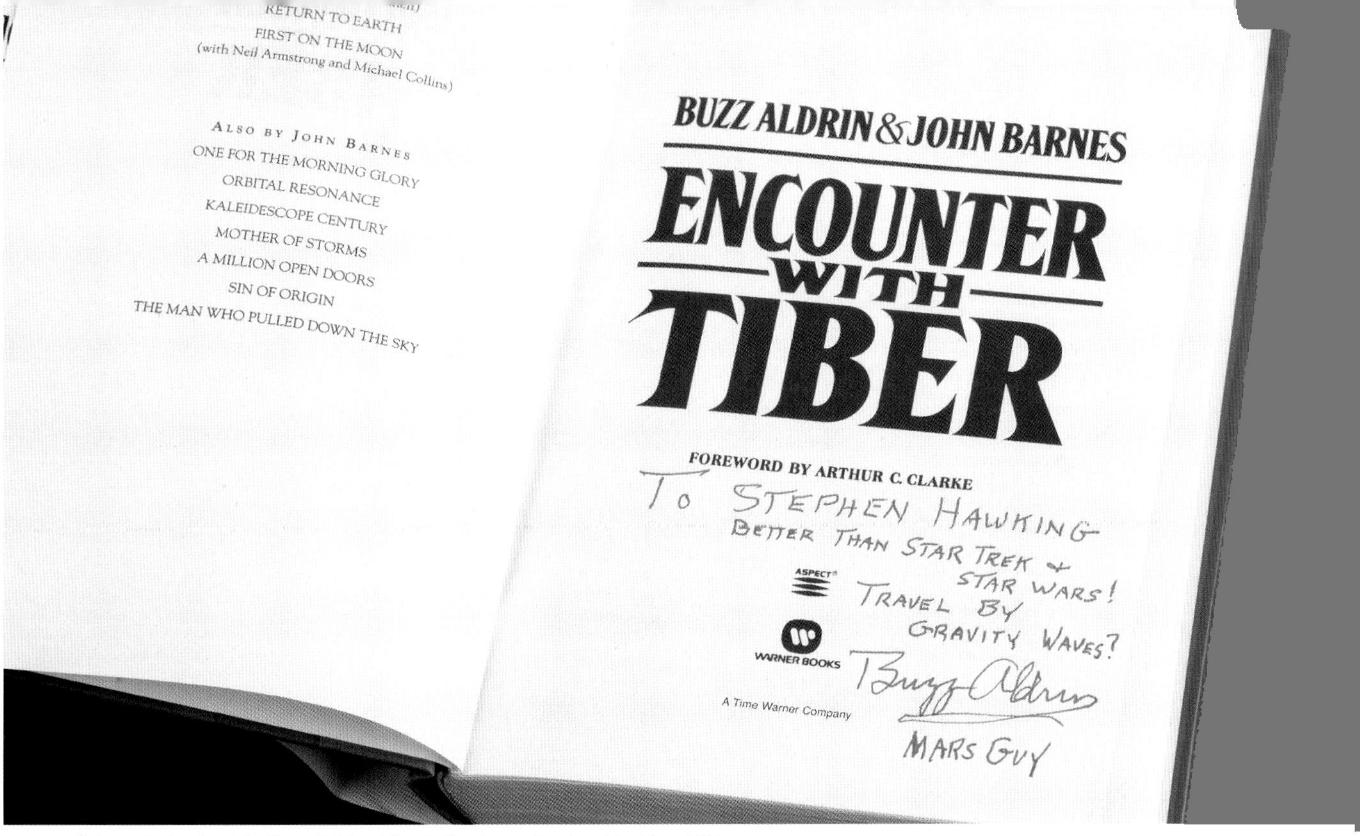
RETURN TO EARTH
FIRST ON THE MOON
(with Neil Armstrong and Michael Collins)
ALSO BY JOHN BARNES
ONE FOR THE MORNING GLORY
ORBITAL RESONANCE
KALEIDESCOPE CENTURY
MOTHER OF STORMS
A MILLION OPEN DOORS
SIN OF ORIGIN
THE MAN WHO PULLED DOWN THE SKY
BUZZ ALDRIN & JOHN BARNES
ENCOUNTER WITH TIBER
FOREWORD BY ARTHUR C. CLARKE
TO STEPHEN HAWKING
BETTER THAN STAR TREK + STAR WARS!
TRAVEL BY GRAVITY WAVES?
Buzz Aldrin
MARS GUY
ASPECT
WARNER BOOKS
A Time Warner Company

ABOVE Science fiction book by Buzz Aldrin with a dedication to Stephen Hawking, 1996.

aerodynamic forces. This counts as space. In the 1900s, the Hungarian physicist Theodore von Kármán reckoned that the edge of space was around 50 miles up, or roughly 80,000 metres, but today the Kármán line is set at 100 kilometres or roughly 62 miles above sea level, by the Fédération Aéronautique Internationale, which keeps track of standards and records in astronautics and aeronautics.

TOO LATE FOR HAWKING

Unfortunately for Stephen Hawking, however, the deadline for that Virgin Galactic flight slipped again and again, and it did not take place before his death. The vehicle's first fully crewed test flight to space eventually took place in 2021, taking Branson and five others more than 50 miles above the New Mexico desert.

"Stephen Hawking really wanted to fly to space, and this would have happened, had the Virgin spacecraft not crashed in 2014," commented Juan-Andres Leon, the Science Museum's Hawking curator, referring to the loss of the VSS *Enterprise* during its fourth test flight, when the SpaceShipTwo experimental vehicle broke up and crashed in California's Mojave Desert, killing co-pilot Michael Alsbury, seriously injuring pilot Peter Siebold and scattering debris for miles.

HAWKING RADIATION

Other aspects of Hawking's space obsession chimed with his own research. "Importantly, and less well known, he had benefitted enormously from space flight as a scientist throughout his career," says Leon, "as, for example, the microwave background observations that confirmed his ideas were only possible with space probes. And through the 1970s and 1980s, he hoped that gamma-ray burst detectors in outer space would detect his 'black hole explosions' – evidence of what we call Hawking Radiation."

"This did not happen," Leon goes on, "meaning that those small black holes are much rarer in the universe than he hoped initially. And the detection of Cygnus X-1, the first good evidence of black holes and subject of the famous bet with his friend Kip Thorne, was also only made possible by space-based telescopes."

SEE ALSO:

Hawking Radiation, p.46
Betting on the Black Hole Information Paradox, p.132
Universe on a Beach Ball, p.150
To Pluto, and Beyond, p.186

America's Highest Honour

Although Stephen Hawking was a British citizen, in 2009 he was awarded America's highest civilian honour by the then President, Barack Obama.

Hawking's assistant, Judith Croasdell, recalled the buzz of excitement when White House officials called to give them the news. "At first 'the Hawk' [her nickname for Hawking] pretended not to know what the Medal of Freedom was, but that was just a special private joke."

His daughter, Lucy, can remember that when he was invited to receive the medal, the family was worried about him making the transatlantic journey. "His health had deteriorated quite a lot." She can remember telling him bluntly, "This trip could kill you." The response from her father, she recalled, was typical. "I don't mind dying in the White House provided I met Barack and Michelle first."

AN UNCOMMON EASE OF SPIRIT

Hawking's physical deterioration was unremitting, but, thanks to the care he had in place, it did not stop him from accepting the invitation. Hawking made the trip to become one of the 16 "agents of change" who received the National Medal of Freedom in August of that year. It is striking that few scientists win this medal, let alone foreigners, but by then his cameos in *Star Trek* and *The Simpsons* had elevated Stephen Hawking to a level of celebrity few scientists ever achieve.

His presidential citation read as follows: "Living with a disability and possessing an uncommon ease of spirit, Stephen Hawking's attitude and achievements inspire hope, intellectual curiosity, and respect for the tremendous power of science."

ABOVE Photograph of Stephen Hawking with American President Barack Obama and First Lady Michelle Obama. President Obama presented the Medal of Freedom to Hawking during a ceremony in the East Room of the White House in Washington, 12 August 2009.

ABOVE Microwave oven with Barack Obama campaign sticker.

"From his wheelchair he's led us on a journey to the farthest and strangest reaches of the cosmos," said President Obama. "In so doing, he has stirred our imagination and shown us the power of the human spirit here on Earth."

"From his wheelchair he's led us on a journey to the farthest and strangest reaches of the cosmos," said President Obama. "In so doing, he has stirred our imagination and shown us the power of the human spirit here on Earth."

The museum's curator of physics, Juan-Andres Leon, added that the office's microwave oven "has an election sticker of Barack Obama, which can be seen prominently on many photoshoots". It had been picked up by one of the carers and, according to Judith Croasdell, Hawking "insisted on having it conspicuously displayed in the office".

Hawking's US visit was such a success that, several years later, the White House was still celebrating, releasing a short video Stephen had recorded at the time in which he spoke of how he had always wanted to understand how things work, and how his disease had not prevented him from exploring the universe with his mind.

When Stephen Hawking died in 2018, President Obama used Twitter to share an image of the visit, with Lucy by his side. It carried a simple message for the late cosmologist: "Have fun out there among the stars."

SEE ALSO:

To Boldly Go..., p.126
Hawking on *The Simpsons*, p.142

Will You Be My Valentine?

The bookshelves and windowsills of Stephen Hawking's office were used to display greeting cards, most of which date from the final years of his life. Among them were labels from a valentine sent in 2007, which were previously attached to images of Albert Einstein and Marilyn Monroe. The valentine came from Layla Sarakalo, an artist who appeared in the movie *Star Trek IV: The Voyage Home* in 1986, which eventually led her to connect with Hawking.

Layla had been filmed during a spontaneous unscripted moment in the streets of San Francisco in what has become known as the "Nuclear Wessel" scene. As no one knew her name, for 20 years she was referred to by fans of the franchise as "the Mystery Woman". That is, until Layla returned to America in 2005 from Paris, where she had been living, at which point she was linked to the film and invited to appear at *Star Trek* conventions.

While being interviewed at Paramount Pictures' studio in 2006 for a feature story on the *Star Trek* website, Layla was told that over the 40 years of the franchise she and Stephen Hawking were the only people to have played themselves. "I went home and looked up this Professor Hawking," she recalled, "and decided to write him an email, say hello and introduce myself." To her surprise, he responded.

Layla had been told that Hawking was "a big fan of *Star Trek*", so she started collecting pictures for him, autographed by every *Star Trek* actor she came across at conventions. It was "something he couldn't do", she said,

so I would do it for him. Then, when I came home, I'd take this large collection of *Star Trek* photos and collected memorabilia and send it in elaborately decorated packages to him at his office.

ABOVE Layla Sarakalo (far right) in *Star Trek IV: The Voyage Home*. Sarakalo appeared as an extra in the film on a whim, when she stumbled across the filming in her home town of San Francisco. **OPPOSITE** Valentine's day note, signed by Hawking fan Layla Sarakalo.

Layla was told that over the 40 years of the franchise she and Stephen Hawking were the only people to have played themselves.

These packages were opened by Judith Croasdell, his amazing personal assistant, who would show them to Stephen, placing the things he liked best around his office.

In the office, for example, is one from a 2006 convention that features William Shatner and Leonard Nimoy from the original cast, with Charlie Washburn, the first black Director's Guild of America trainee, who worked on *Star Trek*.

Layla's correspondence included a birthday card with rocket fairings from the launch of the Mars rover *Curiosity*, dated 2013, a birthday card made out of an autographed photo of "Capt. Kirk wrestling the Gorn" and, of course, the valentine. "There was nothing 'romantic' about my friendship with Stephen," she added. "It was

based on *Star Trek*, space and bits of art." Hawking, returned the favour by inspiring Sarakalo to turn her artistic skills to art and physics, creating images with scientists at NASA's Jet Propulsion Laboratory and Caltech.

Layla even visited his Cambridge office in 2007, while at a London convention. She never got to meet her long-distance friend. But she did get to sit on his sofa.

SEE ALSO:

———

To Boldly Go..., p.126
Hawking on *The Simpsons*, p.142
Hawking and *The Big Bang Theory*, p.178

Time Travellers' Party

The most thought-provoking – and witty – manifestation of Stephen Hawking's enduring fascination with time comes in the form of an invitation to a reception for time travellers, which he sent out *after* the event.

The invitation to the party on 28 June 2009, with the spatial coordinates 52° 12' 21" N, 0° 7' 4.7" E, featured in his TV series for the Discovery Channel, *Into the Universe with Stephen Hawking*, which aired the following year. The version shown here is an embellished Victoriana form, created on a hand-pressed printing press in October 2013 by the artist and designer Peter Dean, and sold as a limited print run with the declaration, "In purchasing this print, you could help to prove that time travel to the past is possible."

Hawking loved parties. In the 2004 BBC TV film *Hawking* he announces, through the actor Benedict Cumberbatch's voice-over, that "I like simple experiments ... and champagne." By broadcasting his invitation widely, Hawking hoped to lure futuristic time travellers to the past to enjoy his cocktail party, which included canapés, flutes of champagne and a banner declaring, "Welcome, Time Travellers." He hoped that copies of it would endure for "many thousands of years", to maximize the chances of it being seen by an actual time traveller.

Despite a theatrical countdown, however, no one showed up, to the delight of Hawking. "What a shame," he gleefully declared. "I was hoping a future Ms Universe was going to step through the door."

THE GRANDFATHER PARADOX
Time travel creates a headache for anyone who believes that causes must come before effects. The so-called grandfather paradox points out that, if you could travel back in time to murder your grandfather, you would thereby be making it impossible for yourself to be born

and therefore able to time-travel in the first place. In a 1992 paper, Hawking had playfully suggested a "Chronology Protection Agency" to prevent such time paradoxes to make the universe safe for historians, a nod towards time police and patrols that appear in various works of science fiction. More formally, his "Chronology Protection Conjecture" is a hypothesis which proposed that laws of physics prevent time travel on all but microscopic scales (more on the reasons why later).

But, time travel remains theoretically possible, maintains Lord Rees, the Astronomer Royal, according to the currently accepted laws of physics.

> Stephen was noting that point, and that there must be some extra constraint over and above what is implicit in Einstein's equations. In other words, we needed the Chronology Protection Conjecture because Einstein's theory did not incorporate it.

The concept of time travel is expressed mathematically by what physicists call "closed time-like curves", explains Lord Rees. While the Earth's gravity produces gentle distortions in space-time, with enough gravity one could distort time so much that it loops back on itself. These closed time-like curves ought, at least in theory, to allow us to revisit some past moment. They can be found in some solutions to the equations of general relativity, Einstein's theory of gravity, which he unveiled in 1915.

OPPOSITE Time Travellers invitation, which was only issued after the event. The original invitations were for a documentary in 2009. This version was printed at New North Press, London, on a Victorian "Albion" press in October 2013 by artist and designer Peter Dean.

Hawking loved parties. In the 2004 BBC TV film *Hawking* he announces ... that "I like simple experiments ... and champagne."

You are cordially invited to

A RECEPTION FOR

TIME TRAVELLERS

Hosted by

★ PROFESSOR ★

STEPHEN HAWKING

To be held in the past, at

THE UNIVERSITY OF CAMBRIDGE

Gonville & Caius College, Trinity Street, Cambridge

Location: 52° 12′ 21″ N, 0° 7′ 4.7″ E

12:00 UT 28 JUNE 2009

NO RSVP REQUIRED

WHITE HOLES

One theoretical way to travel in time could be using portals in space and time – or the blend of 'space-time' featuring in Einstein's work – known as wormholes. These were first postulated in 1916, when the Austrian physicist Ludwig Flamm realized that the equations behind general relativity and a black hole must have an equal but opposite solution, a "white hole", which cannot be entered from the outside, although light can escape. Flamm had noticed that the two solutions could be mathematically connected by a conduit of space-time, and that the black hole "entrance" and white hole "exit" could be in various parts of the same universe, even in different universes.

In 1935 Einstein himself had explored these ideas further, along with the American-Israeli physicist Nathan Rosen, and the two achieved a solution known as an Einstein-Rosen bridge, which could pave the way to the possibility of moving colossal distances through space-time, though there were practical problems. The good news for those like Stephen Hawking who doubted the possibility of human time travel is that this connection is not usable by people: to pass from a black to the white hole, you would have to traverse a singularity, which is infinitely small and dense, where the laws of physics break down.

A further, potentially related, problem was identified by Hawking's early collaborator, the British Nobel laureate Sir Roger Penrose. His research suggested that a black hole has an event horizon – the boundary beyond which gravity's inexorable pull allows nothing, not even light, to escape – so it would not be traversable. Nor could an Einstein-Rosen bridge exist for long enough for light to cross from one part of the universe to the other. In effect, gravity slams this interstellar doorway shut.

ABOVE Lisa Randall and Roger Highfield at the Science Museum's 2017 Director's Annual Dinner.

THE MAW OF THE WORMHOLE

In 1987, a way around these practical barriers to time travel came from another future Nobel Prize winner, Kip Thorne. Working with his Caltech colleague Ulvi Yurtsever, along with Michael Morris, Emeritus Professor at the University of Sussex, Thorne speculated that with the help of quantum theory – the bizarre theory that governs the subatomic world in terms of probabilities, not certainties – it might be possible to travel between various places and times through wormholes. For the maw of the wormhole to be held wide open, and the event horizon to be banished, Thorne and colleagues calculated that its throat would have to be filled with "exotic matter", or some form of field that, because of quantum fluctuations, could exert negative pressure or negative energy associated with antigravity. As a bonus, they showed that you can time travel if you take one of the two wormhole entrances and start to shake it, accelerating to near light speed and then returning to its original location.

However, finding enough of this exotic matter to prop open the wormhole proved a challenge. In 2017, Ping Gao and Daniel Jafferis, both then at Harvard University, and Aron Wall, then at the Institute for Advanced Study in Princeton, reported a way to hold open wormholes with quantum entanglement – a kind of long-distance connection between quantum entities. The known laws of physics suggested that this solution only worked for microscopic wormholes, allowing passage through only for a short time.

TRAVERSABLE WORMHOLES

But then, once again, a closer examination of the physics provided wannabe time travellers with renewed hope. Juan Maldacena of the Institute for Advanced Study and Alexey Milekhin, who was based at that time at Princeton University, published a paper entitled "Humanly traversable wormholes" in the journal *Physical Review D*. They calculated that traversable wormholes would look to us a bit like middling-sized black holes, and could in theory be traversed by future partygoers by adopting a model of a higher-dimensional universe. (Such a cosmos was postulated in 1999 by Lisa Randall and Raman Sundrum in order to understand why some forces are so strong compared to gravity, and which would also predict the existence of dark matter, a source of gravity.) The mouth of these wormholes would be around the size of the Earth. Forces exerted on any temporal travellers would be around 20 times those of our gravity, which is high but survivable by any space-time-surfing humans. Travel times are about 20,000 years, as seen from the outside. However, from the perspective of the traveller, the trip would last just a few seconds, owing to the large distortion of space-time.

Here, then, is one of the consequences of Einstein's theory of gravity, general relativity: that time would tick more slowly for the person inside the wormhole. "As you go inside, you are accelerated to a velocity very close to speed of light," said Milekhin, "and one nice thing about this is that wormhole travel is environmentally friendly. Because it's the force of gravity which accelerates and then decelerates your ship, you wouldn't need to use any fuel."

But, aside from setting up the wormhole in the first place, problems remain. Maldacena and Milekhin's more recent work suggested that a wormhole must be exceptionally clean before it can be traversed by a time traveller. "In our universe, remnants of the Big Bang, the cosmic microwave background radiation, fill the space around us," Milekhin told me.

> But when they go inside the wormhole, they acquire a very high energy so, if they collide with your spaceship, it will be a problem. Inside a wormhole, even this tiny radiation gets so boosted it can burn your spaceship.

And, unlike the wormholes devised by Kip Thorne and his colleagues, known to aficionados as "shortcut" wormholes, the one devised by Maldacena and Milekhin is a so-called "longcut" wormhole – "Imagine something like a long and convoluted pneumatic tube," said Milekhin – and, unlike shortcut wormholes, does not allow time travel, at least not according to our current interpretation of theory.

Understanding time is fundamental to understanding physics, and Stephen Hawking had himself declared that asking whether time travel is possible is a "very serious question" that can still be approached scientifically. In his autobiographical *My Brief History*, however, he admits that few scientists were willing to put their careers on the line for a subject considered "unserious and politically incorrect". And, as he pointed out in his last and posthumous book, *Brief Answers to the Big Questions*, "If one made a research grant application to work on time travel it would be dismissed immediately."

BEHIND THE CHRONOLOGY PROTECTION CONJECTURE

Hawking's faith in his Chronology Protection Conjecture – to stop temporal travellers from muddling cause and effect – rested in part on his efforts to create a grander "theory of everything" by blending general relativity with quantum mechanics. This is still a work in progress, but other scientists have taken up the mantle after Hawking's death and believe time travel can be banned because, even though a wormhole would allow for closed time-like curves, quantum fluctuations would build up to drive the

Understanding time is fundamental to understanding physics, and Stephen Hawking had himself declared that asking whether time travel is possible "is a very serious question" that can still be approached scientifically.

energy density to infinity in the region of the wormholes, destroying them as a result.

The reason for this is based on taking one of the two "mouths" of the wormhole on a round-trip journey at relativistic speed to create a time difference between it and the other mouth. Now a particle or wave could enter the maw of one wormhole, exit the second at an earlier time, then travel once more though the wormhole to repeat the process in a potentially infinite number of loops through the same regions of space-time, building up to wreck the wormhole. Though it was uncertain whether quantum-gravitational effects might prevent the energy density from growing large enough to destroy the wormhole, Hawking conjectured that not only would the pile-up of vacuum fluctuations still succeed in destroying the wormhole in quantum gravity, but the laws of physics would also prevent any type of time machine from forming. This is the Chronology Protection Conjecture.

Despite the no-shows to Stephen Hawking's time travellers' party, a definitive answer to whether time travel is possible, and in what circumstances, will only come with a full theory of quantum gravity. Until then, the subject will sit in the twilight zone between science fact and science fiction. No surprise, then, that Kip Thorne, who over decades of friendship had made various bets with Hawking regarding black-hole physics, acted as a scientific adviser on the science fiction blockbusters *Contact* and *Interstellar*, both of which considered to dramatic effect the possibility of space-time travel through wormholes.

Stephen Hawking, You Are Wrong!

No doubt, it tickled Hawking to put *Stephen Hawking You are Wrong* on prominent display but the title does pose an interesting question about what it actually means to be right or wrong in cosmology.

Stephen Hawking's office contains not one but two copies of this apparently self-published book, written in Hebrew, by Yair Danel. The book's cover talks about the need for creative thinking to deal with the energy and climate crisis, claiming that the answers will not come from the academic and scholarly world which, he argued, are impervious to original ideas.

This particular copy also contains a note from the author, in English, about how the book, his attempt to discover "the truth", had been sent to 50 "powerful people".

There is indeed tension in the world of science between being sceptical and suspicious of mavericks and being creative and open-minded. While never accepting truths that have been handed down on tablets of stone by authority figures, scientific creativity is still constrained by peer review, committees and groupthink.

In all, Hawking's office contained more than 300 books, located on the shelves above the kitchen counter but, when it comes to these two, "they were located with their spines prominently readable, to amuse visitors," commented Juan-Andres Leon, curator of physics and the Hawking office.

No doubt, it tickled Hawking to put *Stephen Hawking, You are Wrong!* on prominent display, but the title does pose an interesting question about what it actually means to be right or wrong in cosmology.

FALSIFICATION

The influential philosopher Karl Popper mulled this question over in January 1921, when he was 18 and listening to a public lecture by Albert Einstein in Vienna's Concert Hall. Although Einstein's theory of gravity, general relativity, "was quite beyond my understanding", he was struck by how Einstein insisted that if there were conclusive evidence against a theory such as his own, it had to be abandoned.

In his autobiography, Popper concluded: "that the scientific attitude was the critical attitude, which did not look for verifications but for crucial tests: tests which could refute the theory tested, though they could never establish it." The implication was that science should focus on trying to falsify a theory, rather than relying on the accumulation of positive inductive evidence. However, as Stephen Hawking himself pointed out, in practice, scientists are reluctant to give up a theory in which they have invested a lot of time and effort. "They usually start by questioning the accuracy of the observations. If that fails, they try to modify the theory in an ad hoc manner."

OPPOSITE Book in Hebrew criticising Stephen Hawking, which contained a personal message from the author.

IT'S NOT EVEN WRONG

Moreover, many would argue that Einstein did not show Newton was wrong when he unveiled his new theory of gravity in 1915: his theory of general relativity had wider applicability and offered deeper insights than Newton's.

Similarly, the progress in understanding the Big Bang back to a nanosecond since the 1960s has been a story of successive refinements, although when it comes to the very beginning, there are many mutually incompatible theories, some of which will be proved wrong.

There is also the famous put-down attributed to the Austrian theoretical physicist and quantum pioneer Wolfgang Pauli (1900–1958): "This isn't right", he is supposed to have said of a student's paper. "It's not even wrong."

When it comes to cosmology, Pauli's phrase is often used to refer to theories or ideas that are so speculative that there would be no way ever to know whether they are right or wrong.

The American theoretical physicist Peter Woit is, for example, well known for asserting that string theory, which for decades has been the leading candidate for a unified theory of physics, is so flawed that it is "not even wrong".

As another example, the theory of inflation predicts not one universe but a multiverse, where everything that can physically happen does indeed happen, and an infinite number of times. Proving the reality of the multiverse seems to lie beyond the reach of the evidence we can muster in our own universe.

In practice, however, it depends on how one defines what we mean by evidence: scientists still take seriously what Einstein's theory says about the inside of black holes – which is, in principle, unobservable from outside – because his theory has been vindicated in many contexts where we are able to make observations.

Ultimately, as the presence of these books attests, Hawking was relaxed about being wrong. For his professional life he made public bets on fundamental questions of physics and he felt no shame about the losses. Rather, he revelled in them because he believed advances occur when scientists are shown to be wrong.

SEE ALSO:
———

Hawking's PhD Thesis, p.34
On the Shoulders of Lucasian Giants, p.62
Not the New Einstein?, p.66
Universe on a Beach Ball, p.150

Hawking and
The Big Bang Theory

It was perhaps inevitable that the world's most famous scientist, who had changed the way we view the Big Bang of creation, would appear in what was then the biggest sitcom on the planet, *The Big Bang Theory*.

At the heart of the show are the antics of physics nerds Sheldon, Leonard, Raj and Howard, played by Jim Parsons, Johnny Galecki, Kunal Nayyar and Simon Helberg, respectively, along with their worldly female foils: Penny (Kaley Cuoco), a waitress, and her co-worker Bernadette (Melissa Rauch), along with neuroscientist Amy (Mayim Bialik).

In one sense it is unsurprising that Hawking was happy to appear on the show. Bill Prady, who created *The Big Bang Theory* with Chuck Lorre, said it was important to them that, as much as was feasible with a half-hour comedy, they depicted research as accurately as possible, or at the very least "did not upset scientists". To this end, they drew on the advice of astrophysicist David Saltzberg of UCLA, and encouraged well-known scientists to visit, and even appear on the set. The sitcom was even reviewed favourably in the heavyweight journal *Science*.

As for how Stephen Hawking ended up on *The Big Bang Theory*, Prady recalled how they had spotted an interview with him in the magazine *TV Guide* in Canada, where the great physicist was asked if he would ever consider appearing. He would, he replied, but had never been approached. "There was a scramble in the office that day to ask him," recalled Prady.

THE GREATEST CONTRIBUTION TO WORLD CULTURE

They eventually managed to get hold of Prady's "dream guest star" after Saltzberg approached Sean Carroll, a theoretical physicist and film consultant at the California Institute of Technology, who in turn put them in touch with Hawking's friend Kip Thorne. Carroll would joke that this was "perhaps my greatest contribution to world culture".

On 9 March 2012, as a consequence of this daisy chain of discussions, Bill Prady announced on Twitter that a "super-secret, super-cool guest star" would appear, after they had filmed his cameo during his residency at Caltech. This was Hawking's most substantial appearance on the show, in an episode that aired that year called "The Hawking Excitation".

The plot was based, as in real life, around a visit to Pasadena to lecture at Caltech. Though Hawking's dialogue was pre-programmed into his speech synthesizer, he wanted to trigger his lines by himself – he wanted to "act", according to Chuck Lorre.

In the episode, Jim Parsons' character, Sheldon, who believes Hawking is "perhaps my only intellectual equal", undergoes a series of gruelling tasks to meet his hero, so that he can ask him to read a paper that "came to me one morning in the shower" on how the Higgs boson particle is a black hole accelerating backwards through time.

"You clearly have a brilliant mind," Hawking declares before giving Sheldon the bad news. "Too bad it's wrong. You made an arithmetic mistake on page two. It was quite a boner [stupid error]."

Bill Prady, who created *The Big Bang Theory* with Chuck Lorre, said it was important to them that, as much as was feasible with a half-hour comedy, they depicted research as accurately as possible, or at the very least "did not upset scientists".

Dear Professor Hawking

...SARCASM
IS MY GIFT TO YOU!
HAPPY BIRTHDAY

With lots of love
& best wishes,
Prakash & Dina
xx

P.S Happy New Year

ABOVE Birthday card featuring *The Big Bang Theory* cast, signed by Hawking fans Prakash D. Nayee and Dina Nayee.

The episode included a fake message from Hawking, impersonated by Howard, left on Sheldon's voicemail: "I wish to discuss your theories of black holes. Meet me at the Randy's Donut by the airport at two a.m."

THE WHEELCHAIR DUDE WHO INVENTED TIME

After filming, Stephen Hawking asked to attend a rehearsal, where Penny referred to Hawking as "the wheelchair dude who invented time" and Howard did his Hawking impersonation. Both Cuoco and Helberg shouted "sorry" as Hawking looked on, recalled Prady, but "he loved it."

Earlier that same year Hawking had missed his seventieth birthday parties because of illness. Now, *The Big Bang Theory* had proved such a happy experience that Prady would join him at a delayed birthday celebration in Caltech and lunch with him.

When Hawking ate, his voice synthesizer would be randomly triggered. Prady suggested that it ought instead to have a pre-programmed repertoire of mealtime quips, from "Mmm, delicious" to "Can I have the recipe?" Hawking responded with "a big smile", Prady recalled, adding that "his assistant told me he was thinking of doing it."

"We would continue to have him on the show, which was just delightful," Prady went on. In one scene, for example, the distinguished cosmologist delivered the following jibe: "What do Sheldon Cooper and a black hole have in common? They both suck. Neener neener." All in all, "it was a remarkable, wonderful experience," said Prady.

The admiration the show felt towards Hawking was reciprocated. On 5 February 2015, when Hawking made another appearance, in the episode "The Troll Manifestation", he posted the following message on Facebook: "If there is any group of people that I'd say have a good shot at cracking my Theory of Everything, it is certainly *The Big Bang Theory*. I wish them luck tonight."

CAKES AND CURRIES

The Big Bang Theory birthday card shown here was sent to Hawking by Dina Nayee, who works at Cambridge University, and Prakash D. Nayee, a senior research scientist at Abzena in Cambridge, fans of the show and the world-famous physicist alike.

They had learned from an article that Stephen Hawking lived on a gluten-free diet, as did Prakash, who had been diagnosed with coeliac disease. "Dina bakes an awesome gluten free Victoria sponge cake and she desperately wanted to bake one for the Prof," recalled Prakash.

Dina approached Hawking in January 2014 with her offer and the couple were invited over to present the cake to him, which was decorated by Dina and into which Prakash crashed a toy flying saucer.

"Much to our delight we were invited to stay and have tea and some of the cake with the Prof and his team," he said. More cakes and curries (hot, Hawking's preference) followed, along with a small Hindu deity of Hanuman, a symbol of strength and energy, now also part of the museum's collection. "Prof had a great sense of humour," said Prakash, "hence *The Big Bang Theory* card and flying saucer crash".

SEE ALSO:

Hawking's Voice, p.90
To Boldly Go..., p.126
Will You Be My Valentine?, p.170

TOP LEFT Hawking with a birthday cake made for him by Prakash and Dina. **OPPOSITE** A meeting of minds: Stephen Hawking on set with Sheldon Cooper, played by Jim Parsons.

Black Hole Light

As Stephen Hawking worked at his desk, to his left sat a curious spiral-shaped light. This was the "Black Hole Light" created by Mark Champkins, the Science Museum's then Inventor in Residence, as a present to celebrate Hawking's seventieth birthday in January 2012.

I can remember how, when I had commissioned Mark to come up with something smart and stylish to give as a present, he was a little overwhelmed: "I was chuffed to be asked but didn't really know where to start." Or rather, he did know where to start – by consulting curators Ali Boyle and Boris Jardine, who were then working on the museum's birthday exhibit – but he was unsure what would emerge from these early discussions. "They were amazingly helpful," he recalled, "explaining a little about his theories about black holes and his work to unite the field of quantum physics with the cosmological. They showed me some images of his office, and his most prized objects and awards." And, of course, they had discussed how light is drawn by extreme gravity into the maw of a black hole.

Boyle and Jardine pointed him towards objects in the museum that have a particular relevance to his work, notably Geissler tubes, an early version of what we would today call neon lighting, invented in 1857 by the German physicist and glassblower Heinrich Geissler. "Geissler tubes are beautiful!" remarked Champkins. A high voltage is applied at each end of the tube, passing through the gas contained within, whether it be argon, air, mercury vapour or neon. The current causes electrons to dissociate from the gas molecules, creating ions, charged molecules. When the electrons then recombine with the newly created ions, light is emitted. Each gas gives off a different colour.

"Geissler tubes led to two developments that relate to Hawking's work," Champkins went on. "Firstly, to the development of the equipment used to discover the electron – the first sub-atomic particle, which in turn, arguably led to the field of quantum physics." Secondly, the Geissler tubes led to the creation of the Crookes radiometer (also known as a light mill), which is made from a glass bulb from which much of the air has been removed, containing a rotor with several lightweight vanes polished or white on one side, black on the other. As its name implies, the radiometer was invented in 1873 by the chemist Sir William Crookes to detect radiation in the form of light, where the mill rotates faster in more intense light. This, said Champkins, linked with Hawking's "identification of his very own form of radiation, that which escapes from a black hole".

BLACK HOLE LIGHT

Taking his inspiration from this object, Champkins decided to make a "Black Hole Light", using the closest thing available to a Geissler tube – a neon tube. He wanted one fashioned into a spiral, inspired by a model he had seen in Hawking's lab, demonstrating how light is sucked into a black hole like water spiralling into a plug hole.

> I had to track down a glassblower to make it, which wasn't easy. Eventually I found an amazing company in the East End of London who took on the task and made a great job of it. When the head glassblower heard it was a gift for Stephen Hawking he said it was nice to do something worthwhile, as he usually made neon signs that say, "Girls, Girls, Girls".

The result was the Black Hole Light. "I liked the pun," Champkins said, "and how it alludes to Hawking radiation."

SEVENTIETH BIRTHDAY

A bout of illness for Stephen Hawking in early 2012 delayed the museum giving him his seventieth birthday present, as well as a series of planned birthday celebrations. But a few months later, Mark Champkins was able to hand over the Black Hole Light in person. Hawking, who always preferred surprises for presents, replied by simply typing, "Magic".

To the delight of visitors, Professor Hawking asked to be taken on a tour of the museum, which he liked to describe as "one of my favourite places". When it was eventually time for him to leave, he told us how much the experience had improved since his childhood: "The museum is much better than when I used to come in the Forties and Fifties."

OPPOSITE Neon light sculpture to celebrate the discovery of Hawking radiation and to mark his seventieth birthday, by the Science Museum's Inventor in Residence, Mark Champkins, presented to Hawking in 2012.

When the head glassblower heard it was a gift for Stephen Hawking he said it was nice to do something worthwhile, as he usually made neon signs that say, "Girls, Girls, Girls".

Virtual Physicist

This thank-you card was sent to Stephen Hawking in recognition of a remarkable virtual appearance he made in 2015 at the Sydney Opera House.

The technology, a first for the physicist, transported Hawking's image from the Cockroft Building in Cambridge University to the Opera House's Concert Hall to enable him to present a lecture at "An Evening with Stephen Hawking". The feat depended on DVE Telepresence Holographic Live Stage technology, and a high-definition video stream provided by Cisco, the multinational digital communications company. As Hawking remarked, "the idea of being the first person to appear as a hologram ... at the Opera House was too good an offer to refuse."

PEPPER'S GHOST

Strictly speaking, the illusion was not a hologram but a version of Pepper's ghost, an effect named after an English scientist, whereby an image is reflected upon a transparent screen at a 45-degree angle, though in this case they used a "3D light mesh". The high-definition video stream projected a three-dimensional image of Hawking on the mesh so that he could appear with his daughter Lucy and physicist Paul Davies live on stage in the Concert Hall. "It was as though my father was in the room," recalled Lucy.

With an eye on popular culture, Hawking even answered a question about One Direction, the English-

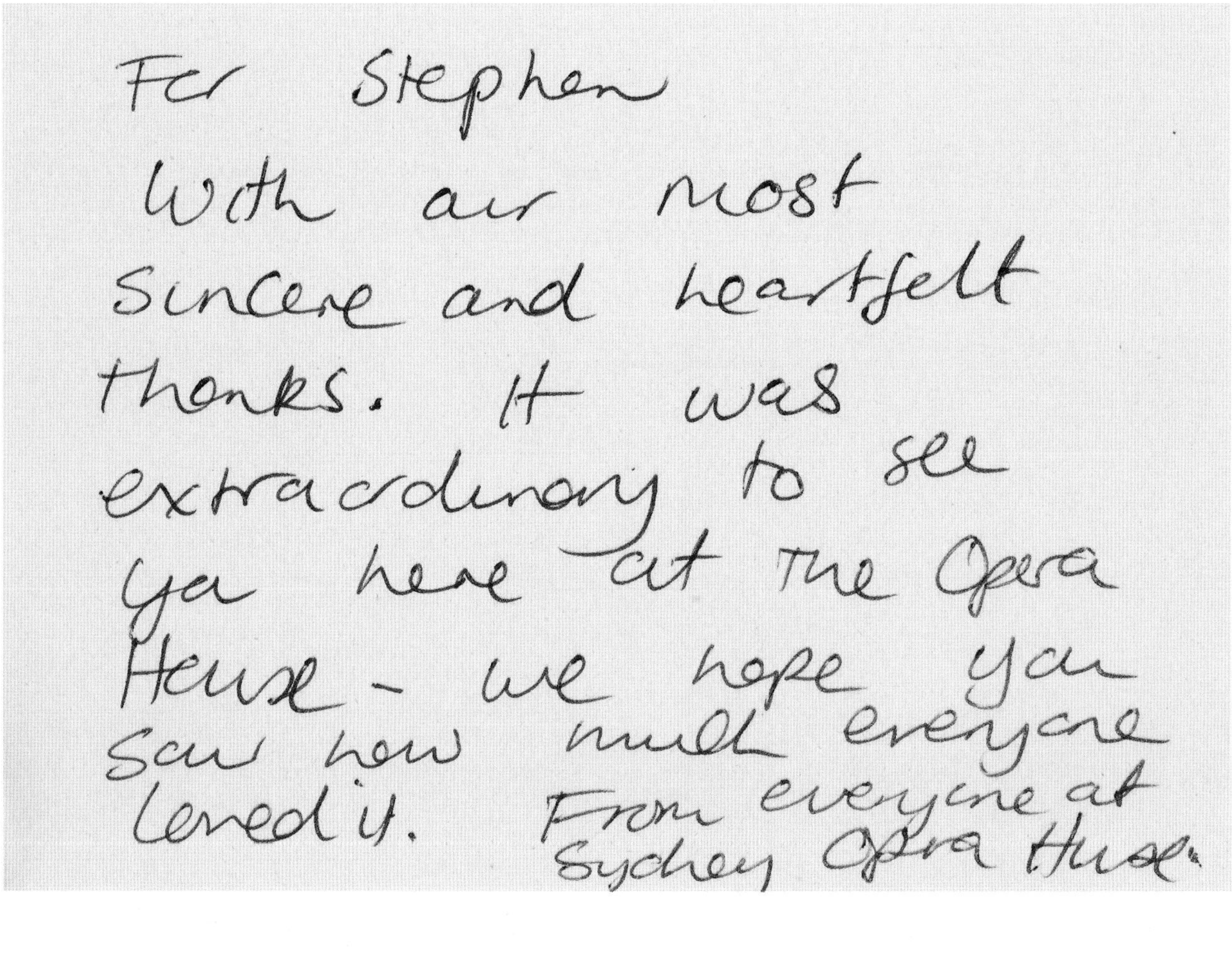

For Stephen

With our most sincere and heartfelt thanks. It was extraordinary to see you here at the Opera House – we hope you saw how much everyone loved it. From everyone at Sydney Opera House.

Maureen Short

Irish boy band, when someone asked about the cosmological consequences of Zayn leaving the group, an announcement that had recently broken the hearts of teenagers worldwide. "Finally," he joked, "a question about something important!" He went on, "Play close attention to the study of theoretical physics because one day there may well be proof of multiple universes" – which would mean, of course, that in one universe "Zayn is still in One Direction."

Hawking finished his lecture with a quote from a character from his favourite sci-fi series, *Star Trek*: "Live long and prosper", before adding, "Now, beam me up." And with that, he vanished.

OPPOSITE AND ABOVE Thank-you card from Sydney Opera House for Hawking's appearance as a 'hologram', 2015–2016.

With an eye on popular culture, Hawking even answered a question about One Direction, the English-Irish boy band, when someone asked about the cosmological consequences of Zayn leaving the group, an announcement that had recently broken the hearts of teenagers worldwide.

To Pluto, and Beyond

"We explore because we are human, and we want to know," Hawking said. "I hope that Pluto will help us on that journey."

Stephen Hawking was one of many people to congratulate the team behind NASA's New Horizons mission, the first spacecraft to explore Pluto up close when it carried out a flyby of the dwarf planet and its moons on 14 July 2015. "The revelations of New Horizons may help us to understand better how our solar system was formed," Hawking said in a video posted to Facebook. "I will be watching closely."

At the time of the spacecraft's construction and up to the mission's launch in January 2006, Pluto was considered our Solar System's ninth planet – the Science Museum collection includes an image taken more than 70 years earlier, when it was discovered by the 13-inch Lawrence Lowell Telescope in Arizona. However, with New Horizons already on course, Pluto was demoted to "dwarf planet" status in a decision by the International Astronomical Union that proved controversial: reflecting this, the museum's collections include bumper stickers that read, "Honk if Pluto is still a planet" and "Pluto got what it deserved".

This octagonal card with information about the mission was printed while Pluto was still considered a planet. "I suspect this object would not be in the office had Pluto not been demoted," commented curator Juan-Andres Leon. "Several of the spaceflight-related items in the office refer to accidents (the Space Shuttle *Columbia*, Virgin VSS *Enterprise*) or, in this case, an awkward situation."

Even so, Hawking congratulated the New Horizons team after it hurtled past the icy world of Pluto at a distance of about 7,800 kilometres (4,800 miles). "We explore because we are human, and we want to know," he said. "I hope that Pluto will help us on that journey."

Thanks to detective work by our Senior Documentation Officer Rowena Hartley, we also realized that a model in his office was a 3D printed copy of New Horizons, which had been created after NASA's Ames Research Center released the spacecraft's blueprint in 2014, followed by a web page entitled "Take New Horizons for a Spin and Print Your Own Model!"

SEE ALSO:
Groovy Icon, Serious Science, p.158
Hawking in Space, p.164

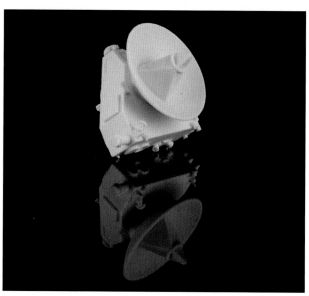

ABOVE A 3D printed model of the "New Horizons" space probe recently identified from the contents of Professor Hawking's office.

ABOVE Octagonal card about the New Horizons space probe built for NASA by the Johns Hopkins University Applied Physics Laboratory and launched in 2006, when Pluto was still classified as a planet.

"The revelations of New Horizons may help us to understand better how our solar system was formed," Hawking said in a video posted to Facebook. "I will be watching closely."

Steaming Inspiration

Given his lifelong interest in trains, it should perhaps come as no surprise that Stephen Hawking's greatest achievement, his discovery of Hawking radiation, owes a debt to the science of steam.

Hawking's discovery that black holes were, in fact, "not so black" demanded not only that he blend general relativity and quantum mechanics, the two great triumphs of twentieth-century physics, but also that he take account of thermodynamics, the great nineteenth-century theory that describes the workings of steam engines, among other things.

Since the steam engine began shaping the modern world through the Industrial Revolution, thermodynamics

has reigned over physics, chemistry, engineering and biology. Thanks to Hawking's work, it seemed that the laws of thermodynamics are also analogous to the laws of black hole mechanics.

Hawking's office contained models of two of the world's most famous and significant steam locomotives: *Flying Scotsman*, the first to officially reach 100 mph in service, and *Mallard*, which still holds the world steam speed record of 126 mph, set on 3 July 1938.

These locomotives are part of the Science Museum Group's collections at the National Railway Museum in York, and the die-cast models of these steam icons confirm Hawking's lifelong interest, which began as a young boy.

ABOVE A "Rail Legends" model of the *Flying Scotsman* steam locomotive, 2013–2014. **OPPOSITE** *Flying Scotsman* at Kings Cross station in 2022 to mark the 170th anniversary of the station and the launch of the locomotive's centenary celebrations.

Thanks to Hawking's work, it seemed that the laws of thermodynamics are also analogous to the laws of black hole mechanics.

In his memoir *My Brief History* (2013), Hawking describes how one of his early memories was getting a train set. "I had a passionate interest in model trains," he wrote, recalling his excitement when his father brought him a present of "an American train, complete with a cowcatcher and a figure-of-eight track".

SATURDAY NIGHT AT THE BALTIMORE
In August 2014, Stephen Hawking became one of the first passengers to travel on a specially adapted wheelchair-accessible carriage, coach no. 4884, hauled by a steam engine on the West Somerset Railway.

At the start of that same year, he bought these model locomotives during a visit to the National Railway Museum in York with his son Tim, whom he had previously brought to the museum as a small child. Jim Lowe, the then Head of Operations, who met Stephen and Tim, recalled how the famous physicist "had a go on the accessible version of the Mallard Simulator ... they were inspired by *Mallard* and thoroughly enjoyed their visit here."

Late that Saturday night, Hawking went out for cocktails at the Biltmore Bar & Grill, housed in a converted church on Swinegate, and posed for photographs with his fans, even venturing out onto the dance floor. "He seemed to have had a very full day out in York," said Lowe.

SEE ALSO:

Hawking's Equation, p.228

Hollywood Hawking

This silver beaker celebrates Stephen Hawking's conquest of Hollywood. Engraved with the formula for Hawking radiation, his greatest achievement, it was given to him in a presentation box by the producers of the biographical film *The Theory of Everything*.

The acclaimed movie was based on *Travelling to Infinity: My Life With Stephen*, the 2007 memoir written by his first wife Jane, which gave a more positive version of their life together than her 1999 memoir, *Music to Move the Stars*. This revision underlined their rapprochement after the turbulent years of their three-decade long marriage, and what he himself described in *My Brief History* as his "passionate and tempestuous" 1995 marriage to Elaine Mason, one of his nurses, which had ended in 2006.

The film takes various dramatic liberties but, as his old friend Lord Rees remarked, it surpassed most biopics in representing the main characters so well that they themselves were happy with the portrayal (even though it understandably omitted and conflated key episodes in his scientific life).

The Theory of Everything conveyed with sensitivity how the need for support (first from a succession of students, but later requiring a team of nurses) strained his marriage to breaking point, especially when augmented by the pressure of his growing celebrity.

Working Title's Eric Fellner, the producer on the movie, remarked on Hawking's "genius" reaction after he first saw the film at a screening: "not everything in it happened ... but all of it was true."

This beaker came with a note of thanks from Fellner and fellow producer Tim Bevan, who have been called the "Batman and Batman" of the British film industry for the dozens of awards they have won. New Zealand-born Bevan and Brit Fellner had their breakthrough in 1994 with the romantic comedy *Four Weddings and a Funeral*, and are also notable for their long-time collaboration with American filmmakers the Coen brothers.

The Hawking biopic garnered numerous awards and nominations, including five Academy Award nominations, and winning Best Actor for Eddie Redmayne. In Lord Rees's opinion Hawking was "superbly impersonated" by Redmayne. "What impressed me was the way Redmayne captured Stephen as he was in the mid-1960s, when he was still able to hobble around with two sticks. I saw him then, but of course Eddie never did."

The actor was among 500 guests invited to Hawking's funeral at the University Church of St Mary the Great in Cambridge, where he gave a reading. Crowds had lined the streets to pay their final respects, and applause broke out as six porters from his former college carried his coffin from the hearse into the church. "We have lost a truly beautiful mind," Redmayne said in a statement, "an astonishing scientist and the funniest man I have ever had the pleasure to meet."

Eric Fellner, the producer on the movie, remarked on Hawking's "genius" reaction after he first saw the film at a screening: "not everything in it happened ... but all of it was true."

SEE ALSO:
Hawking's Voice, p.90

OPPOSITE Silver beaker engraved with Hawking's black hole radiation formula, presented by the producers of the film *The Theory of Everything*.

Comic Relief

Hawking made no secret of his fears that AI or thinking machines could one day take charge, so it is ironic that in one TV sketch he morphed into a super-sized robot that vaporized a couple of well-known comedians, albeit to raise money for charity.

Comic Relief, a charity founded in 1985 in response to famine in Ethiopia, used British comedians to make the public laugh, raising millions of pounds to help people around the world, notably during a telethon on what became known as Red Nose Day.

LITTLE BRITAIN

In 2015, Hawking worked with the comedian David Walliams, who was at the time best known for *Little Britain*, a comedy sketch series based on exaggerated parodies of British people from various walks of life, in a nod towards the pejorative term "Little Englander". The series has not aged well, because of its depictions of minority groups and what were once considered playful jokes at the expense of stereotypes. Nevertheless, Hawking was clearly a fan of the series in its heyday, since he agreed to lend his celebrity to the Red Nose Day sketch for Comic Relief, which was filmed at St John's College, Cambridge.

HA – HA – HA!

Saying on Facebook that he "loved the script", Hawking took up the role usually filled by Andy Pipkin, one of the recurring characters in *Little Britain* played by Matt Lucas: Pipkin falsely presented himself as needing a wheelchair in order to gain the attention of his carer, Lou Todd, played by David Walliams. Walliams reprised the role of Lou and took charge of Hawking. With Lou patronizing him as "Stevie", and assailed on a second front by a creationist and sanctimonious nun played by the comedian Catherine Tate, Hawking sees his wheelchair morph into a Transformer to destroy his persecutors. Laughing "ha – ha – ha!", the professor reverts back to his familiar form and role, heading off to give a lecture "on the origins of the universe".

OPPOSITE Photograph from *Little Britain* sketch featuring Hawking and David Walliams, 2015.

With Lou patronizing him as "Stevie", and assailed on a second front by a creationist and sanctimonious nun played by the comedian Catherine Tate, Hawking sees his wheelchair morph into a Transformer to destroy his persecutors.

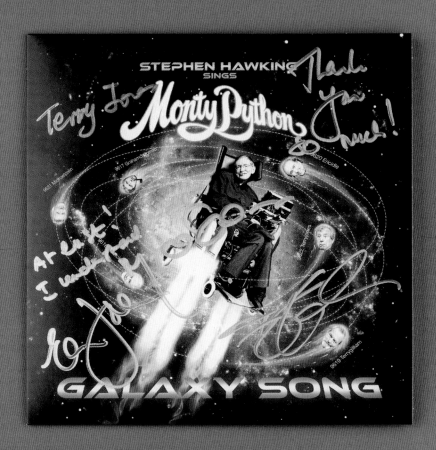

Hawking Sings Monty Python

Stephen Hawking not only tried to unite disparate theories of physics, he also attempted to bridge the worlds of science and comedy. His single of a famous Monty Python song, an autographed and framed copy of which he kept in his office, is a perfect example.

Written by Python Eric Idle and Python collaborator John Du Prez, the "Galaxy Song" is dotted with cosmic facts and astronomical quantities, and first appeared in the 1983 film Monty Python's film *The Meaning of Life*.

Decades later, in 2014, Hawking's version of the song was performed in the stage show *Monty Python Live (Mostly)*, followed by a clip of the TV physics professor and popularizer Brian Cox in the grounds of King's College Cambridge. The two had been persuaded to take part in the sketch by Idle, a friend of Cox.

After he berates the song for its scientific inaccuracies, Cox is knocked over by Hawking racing past in his wheelchair while he is discussing the theory of inflation. "I think you are being pedantic," declares Hawking.

Cox remembered how, during the drive back from Cambridge, he and Eric Idle agreed it was "the most Pythonesque thing you can imagine – we spent a whole day with Stephen and we did not talk about the universe, physics, cosmology or black holes, but being stupid and getting run over for the "Galaxy Song". We had such a fun time."

The song was subsequently released as a single on 13 April 2015 by Hawking, who said in a Facebook post a few days later that "the opportunity to run over Professor Brian Cox and add to my musical resumé were chances I just couldn't pass up."

Hawking and the song also appeared in a web game, Asteroids, on the Monty Python website: "A rocket-propelled, heavily armed Stephen Hawking … must shoot and destroy the Monty Python asteroids (represented by the heads of each of the Pythons) before they collide with him. Points are awarded for shooting all Pythons, wiping out all the bearded Pythons, taking down Brian Cox and zapping both Terrys."

OPPOSITE A framed vinyl record of the "Galaxy Song", 1,000 limited-edition copies were pressed as part of Record Store Day, 2015.
TOP RIGHT A signed Monty Python cartoon: "To Stephen with many thanks", 2015.

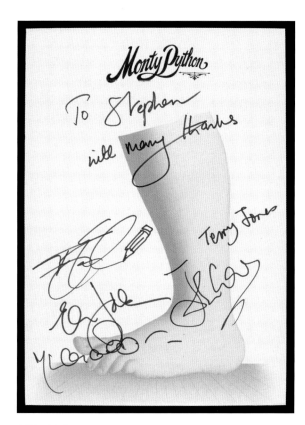

SEE ALSO:

Universe on a Beach Ball, p.150

The Butterfly Effect

Some have likened Stephen Hawking to a "butterfly mind in a diving-bell body", a reference to Jean-Dominique Bauby's memoir about locked-in-syndrome, which is marked by complete paralysis.

This image captures Hawking's encounter with a real-life butterfly, the American opera singer Rena Harms. She had just played the geisha Cio-Cio-San, the Madama Butterfly in the eponymous opera about unrequited love by Puccini, one of Hawking's favourite composers. Her "*Un bel dì, vedremo*" ("One fine day we shall see") is the opera's most famous aria and one of the most popular works in the soprano repertoire.

The picture of their meeting was taken after a performance of Anthony Minghella's production of *Madame Butterfly* by the English National Opera at the London Coliseum on 29 June 2016. Stephen Hawking had sent a message backstage at intermission that he was enjoying the show immensely, and gave his best wishes to the cast.

"After the performance we happened upon each other as I was leaving the stage door," recalled Rena Harms,

> and he requested that we take that picture. I was totally star-struck. He told me (through his friend Gary Kahn, an opera guide publisher) that I was his favourite *Butterfly* he had ever seen. It was truly a once-in-a-lifetime encounter, and I am tickled that he had the photo of us in his office.

SEE ALSO:
———————

Please Please Me, p.122

BELOW Print-out of a photograph of Stephen Hawking with opera singer Rena Hams, 2016.

One Cosmologist, Many Causes

One by-product of Stephen Hawking's global fame was that anything he said attracted a lot of attention.

This was, no doubt, on the mind of the Swedish Crown Princess Victoria, Duchess of Västergötland and heir apparent to the Swedish throne, and Prince Daniel, Duke of Västergötland, a former personal trainer and gym owner, when they visited him in Cambridge late in 2016. The meeting was organized by their Generation Pep charity, a non-profit organization working to give children and young people the opportunity and the will to enjoy active and healthy lives, part of the Swedish Crown Princess Couple's Foundation.

"Today too many people die from complications related to overweight and obesity. We eat too much and move too little," Hawking said in a subsequent "pep talk" video to encourage a healthier lifestyle, created for the charity for use on Swedish television and social media. We need more physical activity and a change in diet, he continued: "It's not rocket science."

The Swedish creative agency Forsman & Bodenfors said that to push home the message about the dangers of being sedentary and expanding waistlines "we reached out to Professor Stephen Hawking, by many considered the smartest man alive, and according to us, someone who truly has a say on this matter."

But this venture outside Hawking's comfort space of cosmology lacked both authenticity and insight: the global problem of obesity, which increases the risk of several debilitating and deadly diseases, has proved much more difficult and intractable than much of rocket science.

"Stephen was far from being the archetypal unworldy or nerdish scientist," commented his friend, Lord Rees. "He had robust common sense, and was ready to express forceful political opinions." However, a downside "was that his comments attracted exaggerated attention, even on topics where he had no special expertise – for instance philosophy, or the dangers from aliens or from intelligent machines."

And, added Lord Rees, in some media events "his 'script' was written by the promoters of causes about which he may have been ambivalent."

ABOVE Photograph from a visit to Hawking's office by Swedish Crown Princess Victoria and Prince Daniel, Duke of Västergötland, 2016.

3
Later Years
and Legacy

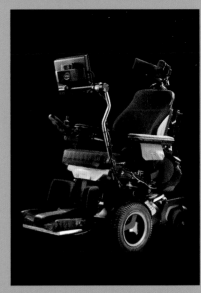

THE ROBERT A. HEINLEIN MEMORIAL AWARD
PRESENTED TO
DR. STEPHEN W. HAWKING
BY THE CHAPTERS AND MEMBERS OF
THE NATIONAL SPACE SOCIETY
AT THE 31ST INTERNATIONAL SPACE DEVELOPMENT CONFERENCE
HELD AT WASHINGTON, DC, MAY 24 - 28, 2012
FOR LIFETIME ACHIEVEMENT IN PROMOTING THE GOAL OF
A FREE SPACEFARING CIVILIZATION

A World of Souvenirs

In "No Boundaries", the thirteenth and final chapter of his memoir, one gets a vivid insight into why Stephen Hawking never complained about his lot: "I believe that disabled people should concentrate on things that their handicap doesn't prevent them from doing and not regret those they can't do."

One can sense his pride when he goes on to say that he travelled widely. His graduate assistant at the department had various suitcases to get him and his equipment – notably his voice synthesizers – safely to every destination, and Hawking's office was decorated with gifts from foreign visitors, along with souvenirs from his various travels around the world.

Top of the list of the excursions he mentions in his memoir were the seven visits he made to the former Soviet Union, mostly to see scientists who were forbidden from visiting the West. "After the end of the Soviet Union in 1990, many of the best scientists left for the West," he explained, "so I have not been to Russia since then." At the end of his first Soviet venture, while travelling in a student group, Hawking was detained for helping to smuggle Russian-language Bibles into the country. This was somewhat incongruous, given that his only religion was science but, as he explained, one member of his student party was a Baptist and asked for help to distribute the bibles: "We managed this undetected, but by the time we were on our way out the authorities had discovered what we had done… however, to charge us with smuggling Bibles would have caused an international incident and unfavourable publicity, so they let us go after a few hours."

RULER IN A SEDAN CHAIR

Hawking's travels ranged from the Forbidden City in China, where he was carried, like a ruler in a sedan chair, by four Chinese wrestlers, to Antarctica in 1997, which he often mentioned in talks about the Hartle-Hawking "no-boundary" proposal, in which he would talk about how "there is nothing south of the South Pole".

These forays are all the more remarkable considering not only Hawking's limited mobility, but also his increasingly fragile health. "He continued to be an inveterate traveller, remarked his friend Lord Rees in

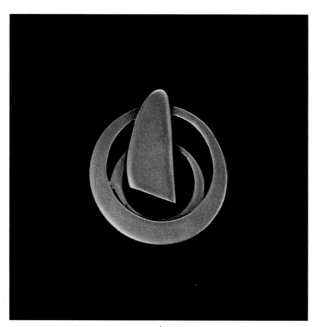

ABOVE A lapel badge from Spain's *Órbita Laika* science show.

ABOVE A Chicago Blackhawks hockey puck.

ABOVE Hawking at the Chilean base in Antarctica, 1997.

ABOVE Stephen's travels required an entourage and array of travel cases containing mobility and communications equipment spares.

his appreciation of Hawking's life, "despite attempts to curb this as his respiration weakened".

AT THE ICE-HOCKEY FACE-OFF

Giving lectures or attending conferences were often the primary reason for these jaunts, but they were certainly not his only motivation. He also showed a genuine desire to both explore and participate: he had always wanted to drop an ice hockey puck at a face-off, and the one among his souvenirs came from the Chicago Blackhawks offering him just such an opportunity.

During a visit to Canada, he was undeterred by having to go two miles down a mine shaft to visit an underground laboratory. When he travelled to Israel, he insisted on also visiting the West Bank in neighbouring Palestine. "Newspapers in 2006 showed remarkable pictures of him, in his wheelchair, surrounded by fascinated and curious crowds in Ramallah," said Lord Rees.

All these travels, said Lord Rees, involved an entourage of assistants and carers. "His fame, and the allure of his public appearances, gave him the resources for nursing care, and protected him against the 'Does he take sugar?' type of indignity that the disabled often suffer."

One long-serving PA, Judith Croasdell, said that in the very early days, only two or three carers would accompany Stephen Hawking on his travels, often enduring spartan conditions on the road. Later travels saw double the number of carers, and the use of private jets. Another crucial companion on his travels, she added, was his graduate assistant – in her day individuals such as Tom Pelly, David Pond, Sam Blackburn and Neel Shearer.

Over a single decade, from 2004 to 2014, Judith travelled with Hawking to Spain (twice), Italy (the Vatican), Hong Kong, China, Chile, Germany, South Africa and Switzerland. As she remarked: "Being an experienced traveller made him forget his limitations." While he picked up souvenirs, visitors would bring him gifts from around the world. The dragon hanging over his office door was given to him during his visit to He Fang Ave, Hangzhou, in China, in 2002, while the embroidered pandas were a present from his Chinese publisher, Hunan Science and Technology Press, in 2006.

SEE ALSO:

Hawking and God, p.54
Beyond the No Boundary, p.156

OPPOSITE A screen of pandas gifted to Hawking by the Chinese publishers of *A Brief History of Time*.

While travelling in a student group, Hawking was detained for helping to smuggle Russian-language Bibles into the country. This was somewhat incongruous, given that his only religion was science.

Accessibility Pioneer

ABOVE Couch from Hawking's office which could also be used as a bed for medical treatments.

Stephen Hawking's office was more than the place where he did his research: it also served as a makeshift clinic and dormitory, in a pioneering example of accessibility in the workplace.

Although it was not immediately apparent to visitors, his sofa had cushions stowed underneath so it could be used as a makeshift bed: "If you look at the dimensions of the sofa", said his daughter, Lucy Hawking, "it's long and it's low, so you can get a person stretched out on it." And above it was a hoist, to help manoeuvre him on to it – a relatively recent addition to what she called "one of the first workspaces of a person with a complex disability". The hoist was a safe handling requirement for people with advanced disabilities, though Hawking was not keen on being moved this way and held out against it for some time.

To have a normal day, Stephen Hawking had to "invest a great deal of effort and dedication", said Lucy, even in simply breathing, as his motor neurone disease weakened the muscles that helped him to inhale and exhale.

For some people with MND various fluids and secretions collect in the lungs, and sometimes Stephen needed to lie flat to get treatment to help clear his airways, though at other times "he just needed to have a little snooze, because he wasn't much of a sleeper."

"I don't think people had any concept of how hard life was on a day-to-day basis," Lucy added, "and the enormous amount of effort he had to put in to have a normal day."

To have a normal day, Stephen Hawking had to "invest a great deal of effort and dedication".

"The Most Influential Disabled Person"

Stephen Hawking was perhaps the world's most famous advocate for people living with disabilities.

Here you can see him presenting Rebecca Dann with an award for her inspiring self-portrait in October 2016, after he had helped to judge the "A World of Unfairness" photographic competition organized by Disability Talk, a media platform for disabled issues. Even though Hawking was not well that day, he decided to push on with the presentation in his office, with Rebecca, the managing director of Disability Talk, Chris Jordan, and the Hollywood actress Eileen Grubba, who has been physically disabled since childhood.

Rebecca has a condition called kyphoscoliosis, an abnormal curvature of the spine. In her winning image, she has her back to the camera and looks over her shoulder, a self-portrait from a project entitled "I'm Fine." Hawking had declared her entry "truly inspirational". "It was overwhelming really," said Rebecca, "to know that I was about to speak about my portrait to the most famous person there is, and the most influential disabled person there is."

Hawking's support for the competition was a small part of his legacy as a disability rights activist, spurred by his motor neurone disease. His involvement with the Motor Neurone Disease Association began when the UK charity was founded in 1979. Among its many activities, he supported the "ice bucket challenge" that originated in the United States, backed by the American national non-profit organization the ALS Association, to promote awareness of the disease. Hawking volunteered his own children for an icy inundation, along with the Director of the Science Museum, Ian Blatchford, and in 2014 alone the challenge raised $115 million for the Association. The Stephen Hawking Foundation, which Hawking launched in 2015, also supported work related to MND, in addition to backing research into cosmology, astrophysics and particle physics.

Though Hawking once said that despite "the isolation imposed by my illness, I feel as though my ivory tower is getting taller", he campaigned more widely about the strains within the UK's health system. He often spoke about how he could never have achieved what he did had it not been for the UK's National Health Service, and, in his later years, opposed cuts and chaos in a service that, as he wrote in one newspaper, "saved me".

The power of Hawking's message whenever he spoke out on living with disability was illustrated when he gave a video message in 2014 to an international UNESCO conference in New Delhi, India. "Recently," he told the delegates,

my communication system broke down for three days, and I was shocked by how powerless I felt. I want to speak up for people who live their whole lives in that state ... Please listen to me. I speak for the people you can't hear.

His message left some in the audience in tears.

SEE ALSO:

OPPOSITE Photograph of Stephen Hawking on 25 October 2016 when he presented a prize to Rebecca Dann, right.

With best wishes and thanks to

PROFESSOR STEPHEN HAWKING

DISABILITY ▐**ALK.CO.UK**

Chris Jordan   Rebecca Dann

Hawking's Last Wheelchair

On TV, Stephen Hawking was often shown roaming the furthest reaches of the cosmos in his motorized wheelchair, a frail figure floating free against the backdrop of a star-spangled blackness. Despite the limitations of his body, his mind could reach the most distant corners of the universe or even travel back to the Big Bang and the origins of time, to glimpse secrets hidden from view for billions of years.

From the late 1960s, as his motor neurone disease progressed, Hawking relied on a wheelchair for his mobility. After 1985, when a bout of pneumonia necessitated an emergency tracheostomy to insert a tube in his neck to help him breathe, he was rendered speechless, and his wheelchair was fitted with a voice synthesizer. Subsequently, the image of the wheelchair-bound physicist with a synthetic voice questing for the profound secrets of the cosmos became an inspirational example of the power of mind over matter.

Over the years, Hawking went through a series of wheelchairs, in which he explored his local, limited patch of the universe. In Cambridge "he became a familiar figure navigating his wheelchair around the city's streets," recalled Martin Rees, the Astronomer Royal. I well remember accompanying his motorized wheelchair around the Science Museum, above all during a 2012 visit to see his seventieth birthday exhibition, which drew huge crowds, along with having to write about his motorized escapes, when he could be reckless.

A SPACESUIT

But in one important sense the depictions of Hawking floating around the heavens in his wheelchair were correct. According to Jonathan Wood, his last technical assistant, who was based in the office next door, Hawking's later wheelchairs actually functioned like a spacesuit. They were equipped with a power supply and critical life support and communications systems. As Hawking's needs changed, or the technology improved, his wheelchairs had to be constantly updated, and at any given time two chairs were generally in rotation, to allow him the use of a backup if required.

Here you can see his Permobil F3 powerchair, the last model Hawking used until his death in 2018. Permobil's Chief Information Officer, Olof Hedin, had wanted to donate a sitting version of the company's latest F5 chair, Jonathan Wood recalled, but it turned out to be too heavy, an issue when it came to using lifts, and too big, notably for an August 2015 flight to Sweden to attend a conference and visit the King of Sweden.

For this Swedish trip, Permobil therefore worked with Wood to customize a lighter F3 wheelchair to Hawking's specific needs, and also provided a wheelchair-accessible van. Over dinner at the end of the trip, Hawking said how much he liked the F3 wheelchair, so Permobil gifted it to him.

When it was delivered back to Cambridge Hawking reiterated his long-held wish to be able to drive his own wheelchair again – something he hadn't been able to do since losing the use of his hands. Wood looked into the possibility of Hawking using a chin joystick. "He drove across the coffee room at DAMTP," Wood recalled, "but we weren't able to make it work reliably for him". Then Wood

OPPOSITE Hawking's Permobil model F3 Corpus wheelchair with communication system, built around 2016.

The image of the wheelchair-bound physicist with a synthetic voice questing for the profound secrets of the cosmos became an inspirational example of the power of mind over matter.

I thought my time was up at the age of 59·97, says Hawking

By ROGER HIGHFIELD
SCIENCE EDITOR

STEPHEN Hawking revealed yesterday how he "had an argument with a wall" and almost did not make the celebrations for his 60th birthday.

"The wall won," said Prof Hawking, addressing his birthday symposium in Cambridge attended by distinguished academics and celebrities from around the world.

"But Addenbrooke's Hospital did a very good job of putting me back together again."

Prof Hawking, the world's best known cosmologist and author of the best-selling books *A Brief History of Time* and *The Universe in a Nutshell*, was introduced as "one of the greatest time lords of all".

Speaking at the symposium on the future of theoretical physics and cosmology, he described how he had his accident a few days after Christmas at the age of "59·97 years".

His graduate assistant, Neel Shearer, said later: "He was late for a meeting and running on Hawking time, as ever." His Quantum Jazzy wheelchair crashed into a wall and Prof Hawking broke his hip.

He was in hospital for five days, but "as you can see, he is in fine fettle now".

Prof Hawking was born on Jan 8, 1942, in Oxford. He was diagnosed with motor neurone disease when he was 21.

Yesterday, Prof Hawking said he thought that he would not survive to finish his Cambridge doctorate. "Then suddenly, things picked up. My disease wasn't progressing much, and my work all fell into place."

He described his great disappointment at being turned down for the chance to work with the late Sir Fred Hoyle, "the most famous British astronomer of the time".

Hoyle coined the term Big Bang to deride the theory that would eventually replace his own "steady state" theory of the universe.

Prof Hawking gave Hoyle his come-uppance in 1964, when only 22, at a meeting held at the Royal Society, London. During a question session, the young scientist undermined a new theory of gravity that Sir Fred had just unveiled.

"Hoyle was furious," he said yesterday. "However, he later gave me a job, so he didn't harbour a grudge."

Prof Hawking went on to describe his research into black holes, and his calculation that they would give off what is now known as Hawking Radiation.

"I would like this simple formula to be on my tombstone," he said yesterday.

He added: "It has been a glorious time to be alive and doing research in theoretical physics, and I'm happy if I have made a small contribution.

"I want to share my excitement and enthusiasm. There's nothing like the Eureka moment of discovering something that no one knew before. I won't compare it to sex, but it lasts longer."

Come-uppance: Sir Fred Hoyle

❝I would like the formula for black holes to be on my tombstone❞

Time lord: Stephen Hawking arrives at his 60th birthday symposium in Cambridge yesterday

Picture: ROB BODMAN

LEFT *Daily Telegraph* article on Hawking, including how he wished for the formula for black hole radiation to be on his tombstone, January 2001.

explored the feasibility of using a new interface integrated into Hawking's software keyboard that had been developed by the semiconductor chip company Intel.

A second F3, with a more advanced joystick controller, was built in Sweden in preparation for an upcoming trip to America, then shipped out to Nashville, Tennessee, for customization by Permobil's Jeremy Satterfield and Nathan Rose. The finished chair was delivered to Hawking in New York City to coincide with his visit in April 2016 to announce the Breakthrough Starshot Initiative with the Israeli entrepreneur Yuri Milner, a US$100 million programme to develop a fleet of "light sail" spacecraft capable of making the journey to the star Alpha Centauri at a fifth of the speed of light. Hawking also spent time at Harvard University, near Boston, with Professor Andy Strominger, and a black hole workshop by Professor Avi Loeb.

THE BLACK HOLE WHEELCHAIR

To make these trips feasible, a lot was packed into Stephen Hawking's wheelchair. The batteries to power the motor and electronics were under his seat. The left side carried an Intel-branded laptop holder with custom wiring, along with a 12.5-inch Lenovo Yoga 260 tablet. "One of the cool things we did for his F3 wheelchair," Mr Rose added, "was to make it unique by customizing the colour trims with black and white to mimic a black hole. It was a unique F3 Black Hole edition." Mr Satterfield admitted to being "a little starstruck" when he met their celebrity customer. "I've met a lot of users at different stages of ALS [the commonly used term in the US for motor neurone disease], and I even have a family friend who has the slow-progressing type, and I was amazed at how much Professor Hawking could still do." He was struck by how the scientist "ate a NY strip steak and potatoes for dinner".

Famously, having run over Prince Charles's toes at a meeting in 1977, Stephen Hawking is said to have regretted not doing the same to Margaret Thatcher, when she was Prime Minister.

On the back of the F3 wheelchair in the museum collection is a black box that housed Hawking's voice synthesiser. While Hawking used it, the back of the wheelchair also held a ventilation system to help him breathe, and there was a joystick controller and small bag attached to the arm rest. Typically, according to Juan-Andres Leon, the collection's curator, this bag contained sunglasses, tools for maintaining the chair, electronics, medicine-related items and various things Hawking accumulated while out and about.

However, this was more than just a wheelchair. It could also be a weapon. Famously, having run over Prince Charles's toes at a meeting in 1977, Stephen Hawking is said to have regretted not doing the same to Margaret Thatcher, when she was Prime Minister. In a comedy skit, he knocked over the physics popularizer, Brian Cox. Indeed, it was claimed, anyone who annoyed him risked becoming a target, an allegation he responded to with characteristic wit: "A malicious rumour. I'll run over anyone who repeats it."

SCRAPES

Over the years, Stephen Hawking had a number of adventures in his motorized transport. One of the most dramatic came a few days after Christmas 2001, when his Quantum Jazzy wheelchair crashed in Cambridge and toppled over, resulting in Hawking banging his head and breaking his femur. "He was late for a meeting," his graduate assistant, Neel Shearer, subsequently told me, "and running on Hawking time, as ever".

The following January, at a Cambridge event to mark his sixtieth birthday, a symposium entitled "The Future of Theoretical Physics and Cosmology", attended by distinguished academics and celebrities from around the world, he told the gathering how he had had "an argument with a wall" at the age of "59.97 years", and almost did not make his celebrations. "The wall won." Hawking had ended up in hospital for five days, but reassured his audience that "Addenbrooke's Hospital did a very good job of putting me back together again."

The sixtieth birthday celebrations continued with a private party attended by a couple of hundred friends, colleagues and former students in Gonville and Caius College. A Marilyn Monroe impersonator – of course – sang "I Wanna Be Loved by You", and Hawking grinned as the assembled throng sang "Happy Birthday" and celebrated with champagne. By then he'd quit drinking, so could not be accused of drink-driving in charge of a wheelchair. "He no longer has a taste for alcohol," his personal assistant at the time, Karen Sime, told me. " He is much more likely to keep drinking tea all night."

ABOVE Hawking in Cambridge. He first arrived in the city in 1962 as a PhD student and rose to become the Lucasian Professor of Mathematics in 1979.

In all, the Science Museum acquired six of Stephen Hawking's wheelchairs – most of those still in existence – with the rest passing into private ownership. At an auction at Christie's in 2018, one motorized chair, which had a pre-sale estimate of £15,000, went for £296,750, with the proceeds donated to his foundation and the Motor Neurone Disease Association. Made in England by BEC Mobility in around 1988, and upholstered in red and maroon leather, it was thought to be the earliest surviving example of a wheelchair used by Stephen Hawking.

Such an extraordinary price was, of course, testament to the wheelchair having become such an integral part of Stephen Hawking's identity and public persona. "The downside of my celebrity is that I cannot go anywhere in the world without being recognized," he once reflected. "It is not enough for me to wear dark sunglasses and a wig. The wheelchair gives me away."

SEE ALSO:

Decorated Scientist

Some of the most striking objects in Stephen Hawking's office were the awards, prizes and gongs that populated its windowsills and bookshelves. "These trophies often changed location," remarked the Science Museum's Hawking office curator, Juan-Andres Leon,

and also circulated in and out of the office, to his home and even reached external exhibitions, such as one held in the Science Museum to mark his seventieth birthday. They represent a small selection of prizes that he won, and were put on show because of their artistic quality, significance or because they represented causes that he held dear.

This is the Pride of Britain Lifetime Achievement Award Hawking was given in 2016, enhanced by the addition of a pink pom-pom, thanks to Stephen's granddaughter Rose.

For the awards, members of the public were invited by the *Daily Mirror* newspaper to nominate British people who make the world a better place. These typically range from children facing daunting adversity to inspirational campaigners, along with individuals who have displayed awe-inspiring courage.

The nominations were sifted by a celebrity panel, which included the creator of Pride of Britain, Peter Willis of the *Mirror*. "Pride of Britain is about recognizing people that inspire us all," he said, "and Professor Stephen Hawking has done that right across the board."

During the awards, which were televised, Hawking received his award from the then Prime Minister Theresa May, who at the time was negotiating the UK's departure from the EU. He thanked the Prime Minister, and could not resist making a quip about her wrangles with Brussels. "I deal with tough mathematical questions every day, but please don't ask me to help with Brexit."

AI ARMAGEDDON

The GMIC (Global Mobile Internet Conference) Award, 2017, was the most recent trophy in Stephen Hawking's office, and one of the few that he did not receive in person. It was awarded to him when, connected remotely to a massive screen at the China National Convention Center in Beijing, he delivered a keynote speech which warned that "AI could spell the end of the human race." The narrow forms of artificial intelligence developed so far, he conceded, have already proved very useful. But he feared the consequences of creating AI that can match or surpass humans, which would be able to re-design itself at an ever-increasing rate. "Humans, who are limited by slow biological evolution, couldn't compete, and would be superseded."

As one commentator remarked, it was an unusual way to start a technology conference featuring many of China's major AI and robotics firms.

LEFT Windowsill in Hawking's office full of awards and gifts. **OPPOSITE** Pride of Britain Award with added pom-pom by Stephen's granddaughter.

Hawking's Most Lucrative Prize

This receipt from a chocolatier in Geneva was only discovered when Nicola Onions, a Science Museum photographer, was preparing to shoot Stephen Hawking's wheelchair. "It was found under the cushion of the wheelchair that Hawking took to Geneva," commented curator Juan-Andres Leon. "When we took a look at his other Permobil F3 and C350 wheelchairs, we found several items associated with them, from expected ones like sunglasses and lens wipes, to various things that we still don't know much about, including a tiny seashell."

The receipt is the only evidence, among all the objects collected by the museum, of Hawking being awarded the $3 million Fundamental Physics Prize – the most lucrative in science – established in July 2012 by the Russian internet tycoon Yuri Milner. Hawking kept the trophy at home.

The 2013 prize recognized Hawking's "deep contributions to quantum gravity and quantum aspects of the early universe". Flanked by his children Lucy and Tim, at a glittering ceremony in Geneva hosted by the actor Morgan Freeman, Hawking talked about his delight in winning the prize, which recognized the work of scientists, like his own on Hawking radiation, that is difficult or impossible to confirm with an experiment. "I thought my discovery would never be confirmed or recognized," he said. However, he added, there was indirect evidence to prove his theory correct, in the form of the echo of the Big Bang represented by the cosmic microwave background.

ABOVE Receipt for Chocolaterie in Geneva, Switzerland, 2013.

He thanked the Prime Minister, and could not resist making a quip about her wrangles with Brussels. "I deal with tough mathematical questions every day, but please don't ask me to help with Brexit."

Among Stephen Hawking's Other Awards in the Collection Are:

The Courage Trophy, in the form of a bronze winged female, fashioned by R. T. Granford in 1980 and presented in 1989 by the Courage Center, Golden Valley, Minnesota.

The Franklin Medal, 1981, a science award presented from 1915 until 1997 by the Franklin Institute in Philadelphia, Pennsylvania, to many significant figures, from Thomas Edison and Guglielmo Marconi to William Bragg and Albert Einstein. The citation recognized Hawking's "contributions to theory of general relativity and black holes".

The medal features in *A Brief History of Time*, when Stephen Hawking describes how, before speaking in Philadelphia after receiving the medal, he had travelled to Moscow to spread the word about problems with the early thinking around inflation, first proposed in 1980 by Alan Guth, who liked to call it the theory of the "bang" of the Big Bang.

Consejo Superior de Investigaciones Científicas gold medal, 1989, from the CSIC, the Spanish Council for Scientific Research.

The striking 1989 Prince of the Asturias Award for Concord, designed by Joan Miró (1893–1983), awarded to Stephen Hawking for "his transcendental research work into the foundations of space and time", and which appeared in the Science Museum's seventieth birthday exhibit.

2008 Fonseca Prize trophy for science communication, an annual award created that year by the University of Santiago de Compostela and the Consortium of Santiago, named after Alonso III Fonseca, one of the earliest patrons of the university. "If I succeeded in allowing more people to wonder after the beauty of our Universe", Hawking remarked on receiving it, "my mission has been accomplished." The trophy was also displayed at the Science Museum.

The Cosmos Award trophy, designed by B. E. Johnson and awarded by the Planetary Society for Outstanding Public Presentation of Science, given to Hawking in 2010 and also displayed in his seventieth birthday exhibit in the museum.

The Robert A. Heinlein Memorial Award, 2012, named after the author widely recognized as the "dean of science fiction writers" in recognition of his "outstanding and continuing public efforts in support of human space development and settlement".

The Innovation Luminary trophy, 2013, one of the first annual European Awards for Innovation Leadership, presented at a ceremony held at Trinity College, Dublin, and introduced by the President of the EU Commission, José Manuel Barroso.

BBVA Fronteras del Conocimiento Award, 2016. The Foundation Frontiers of Knowledge Awards honour fundamental disciplinary or interdisciplinary advances, which in Hawking's case was connecting quantum theory to the science of the cosmos.

Beacon of Courage and Dedication Award, given to Hawking in 2015 by the Society for Brain Mapping & Therapeutics.

The Sir Arthur Clarke Lifetime Achievement Award, presented by the Arthur C. Clarke Foundation in 2017 to Stephen Hawking in Cambridge to mark the hundredth anniversary of the birth of the renowned writer, futurist and technologist, who in his later years was also confined to a wheelchair, in his case by post-polio syndrome. The video recorded of Stephen Hawking for these centenary celebrations showed many shots of his office, and included a quip about the award being for his lifetime "so far".

The Meaning of Life

For one schoolboy, Stephen Hawking was the obvious person to contact when, in January 2018, his Year 3 English class in Dorset was encouraged to write to a famous individual who had inspired them.

Dillon, in class 3C of Wyke Regis Church of England Junior School in the southwest of England, decided to write to Stephen Hawking to tell him how much he admired him. "I love the talking box that is attached to your brain," he wrote, adding that, when he spotted him in the sitcom *The Big Bang Theory*, "I get excited."

To Dillon's amazement, his school reported, Dillon received a reply from his hero, together with a "fantastic fridge magnet" that read:

REMEMBER TO LOOK UP AT THE STARS AND NOT DOWN AT YOUR FEET. TRY TO MAKE SENSE OF WHAT YOU SEE, AND WONDER ABOUT WHAT MAKES THE UNIVERSE EXIST. BE CURIOUS!

Sound advice, don't you think?

ABOVE Magnets with quote by Stephen Hawking, taken from his address to the 2012 Paralympic Games opening ceremony.

Hawking's Living Legacy

This stack of books documents a series of cosmic scientific adventures undertaken by more than 40 students who worked on their theses with Stephen Hawking over half a century, some of whom later made significant names for themselves in research. Each one marks the culmination of their doctoral work with him on a scientific problem, from black holes, to wormholes, to relativity, to a theory of everything, representing a celebrated school of thought that had grappled with some of the most profound questions of physics.

In Hawking's early years, his PhD students helped take care of him and translate his increasingly hard-to-understand speech. By the time he needed round-the-clock nursing, he still relied on collaborators to develop and complete his high-risk, high-gain research projects. Throughout his career, his students were both the fulcrum of his research and, in some respects, a second family. In 2019, an obituary written by his peers for the Royal Society remarked: "Stephen supervised about 40 graduate students, some of whom later made significant names for

ABOVE The shelf in Professor Hawking's office carrying the theses of many of the doctoral students that he supervised over the years.

themselves." Yet, they added, being a student of his was not easy. "He was known to run his wheelchair over the foot of a student who caused him irritation."

THE SEVENTIES
One of Stephen Hawking's first academic offspring was Gary Gibbons (the very first was Chris Prior, who had studied relativity.) Gibbons had started his research in October 1969 under Hawking's own supervisor, Dennis Sciama, and Stephen Hawking took over at the end of the year when Sciama moved to the University of Oxford. At that time, said Gibbons, Hawking "was in a wheelchair and could also speak in a mumbling sort of way, but it wasn't that difficult to understand." He was still able to drive his three-wheeled Reliant Robin, Gibbons recalled.

Under Hawking's supervision, Gary Gibbons would work on detecting gravitational radiation, ripples in space and time. In autumn 1970, Gibbons visited American institutions to explain their research and thinking, and met up with Stephen in Austin, Texas, where he talked about his area increase theorem for black holes. "I hadn't quite taken on board how serious he was about this gravitational radiation experiment."

During that time, recalled Gibbons, "it was easy enough to communicate with Stephen. You had to listen carefully, helped by his facial expressions. But he could move his face pretty well, and he could move about. He needed assistance with walking, and he was also writing, although his writing was deteriorating as time went on."

They continued to work together but, after Gibbons returned from working at the Max Planck Institute in Munich in 1976, he was struck by Hawking's decline. He was now using a wheelchair and, before it was motorized, needed help getting home to his house in West Road, which was around the corner from Silver Street: at that time streets lacked dropped kerbs and were not easy for wheelchairs to navigate.

At Stephen Hawking's lectures, a student would translate what sounded like groaning or croaking, as Hawking had to talk out of the right-hand side of his mouth and would draw on a blackboard. His students would go on to act as his interpreter for a few more years until in 1985 he lost his voice completely.

ABOVE A lunchtime seminar with Hawking and his students in the old Silver Street offices of DAMTP. Gary Gibbons is seated on the right.
OPPOSITE Hawking at Cambridge University, 1988.

The speech synthesizer made communication more time-consuming and slower, recalled Gibbons, who worked with Hawking into the 1990s. But in some ways, this helped. His illness forced him to take things slowly, thinking things out with great thoroughness, before concisely laying out his ideas. As another student, Thomas Hertog, remarked: "Like the Oracle of Delphi, he had mastered the art of packing a lot into a few words."

THE EIGHTIES
Fay Dowker, now Professor of Theoretical Physics at Imperial College, London, worked with Stephen Hawking from 1987 until 1990. Her recollection that his commitment to physics was absolute chimes with that of his first wife, Jane Hawking, who remarked in a biographical movie that it was "almost as if everything had to be sacrificed to what I call the worship of the goddess of physics." "Everything revolved around the physics," said Dowker. "That's what made it all so great – that time was so important to me." Even better, Hawking was highly sociable, and liked being the life and soul of the party.

Throughout his career, his students were both the fulcrum of his research and, in some respects, a second family.

His illness forced him to take things slowly, thinking things out with great thoroughness, before concisely laying out his ideas.

Dowker had joined him to study wormholes, shortcuts in space-time, as a means of understanding what happens to information that falls into a black hole. Because of Hawking radiation, a black hole will evaporate entirely – and so too will any information about its swallowed contents. This is known as the black hole information paradox. Hawking wanted her to study wormholes as a possible route for information to escape into "baby universes", though Dowker adds that such "umbilical" wormholes are a "different species" from the wormholes being considered for time travel.

Dowker was his first female student. Partly that reflected how, all those decades ago, women were a relative rarity in the physical sciences. "There were very few women, full stop," she said. However, there may have been another reason. Around that time Stephen Hawking started receiving professional care; until then, his graduate students had had to be part-time carers, helping him get to the bathroom, for example. "I think that would have been difficult for a female student."

Fay Dowker could vividly remember how she made quite an impression on the professor when she arrived. "For some reason, and I can't really explain it, I had my hair shaved off. He had the most expressive eyebrows in all of physics," she went on, and on seeing her, "they nearly shot off the top of his head. His facial expression said, 'What kind of student have I taken on?'"

"I see you had a fight with a lawn mower," Hawking typed. A dramatic pause, then: "... and lost."

As they worked together, Dowker would look over his shoulder at his computer screen to anticipate when he would say, "Calculate this." She would write some calculations on the blackboard, "and he would often say, 'No'." This was, she recalled, as empowering as it was terrifying. "He could be quite irascible, but he was very, very approachable. There was none of this, 'I'm the big shot fancy professor, and you're the minion.' He expected his students to play their full part in his research programme. He did not make allowances, that we were just students, and that was exhilarating, really exhilarating."

The setup in the old Silver Street building of the Department of Applied Mathematics and Theoretical Physics was conducive to collaboration: their offices were among eight adjacent to the tea room, the beating heart of DAMTP and in some ways superior to the purpose-built Centre for Mathematical Sciences completed in 2003. Not in every way, however: Hawking could only enter the old building thanks to "Hawking's Ramp", a tatty wooden construction tacked onto the back.

"I remember it vividly – we spent so much time there," said Dowker. The olive-green door to Stephen's office opened into the tea room, "which always felt to me as if it could be underground, due to a lack of natural light. It was the most unromantic place, with Formica tables that we would scribble on, but it was my life for three years. It was all-absorbing, all-consuming. We lived in that tea room, and he would join us for lunch there." There they could puzzle over the great mysteries of the universe as their tea stewed in a big aluminium teapot.

By then, the success of *A Brief History of Time* had seen Hawking's celebrity status soar. "Film stars turned up," said Fay Dowker, "and we never knew when there would be a TV crew too", but his "tender and naïve" students remained oblivious to the fuss. Even their stellar professor took a more askance view of fame, which he dismissed as "being known by more people than you know".

The inner circle with whom he had worked a long time, such as Thomas Hertog and Jim Hartle, were all experts at reading his face.

ABOVE Stephen Hawking with student Chris Hull, Cambridge, 1985.

One year the actress Shirley MacLaine turned up at their Relativity Group Christmas Lunch. She "flounced in an hour late in a leopard-skin outfit to find us sitting round looking nerdy, wearing the paper hats from our Christmas crackers," recalled Fay Dowker. "A surreal scene ensued, with Shirley bending Stephen's ear about her New Age theories. She argued that since Stephen talked about energy, and since she knew the world ran on love energy, they must be talking about the same thing. Occasionally there would be a pause and Stephen would say 'No', and 'Wrong', though this didn't seem to dampen her enthusiasm at all." This was all long before today's social media culture, and such hoopla passed Dowker and her young peers by. "We never took a single photograph."

THE NINETIES

Another of Stephen Hawking's protégés was Marika Taylor, now Pro-Vice-Chancellor and head of the College of Engineering and Physical Sciences at the University of Birmingham, who had been inspired as a student by *A Brief History of Time*, along with a series of Cambridge lectures given in 1994 by Hawking and Penrose.

In 1995, after finishing her fourth year at Cambridge, Taylor was invited in to talk to Hawking about options for her doctorate. "I was quite nervous when I first met him, but soon we were discussing black holes." Theoretical physicists typically give their early PhD students "safe" research projects, and guide them through their calculations but, like Fay Dowker, Taylor found Hawking's students did not have the luxury of a gentle introduction – "he needed us to work on his own high-risk, high-gain projects."

"You could jump in to anticipate what he was going to say with his speech synthesizer," Taylor recalled. "I was hesitant to do this at first but learned that you had to pay close attention to his facial expressions. If he was rolling his eyes, you were clearly not going in the right direction. Sometimes, you would get it right and you could see a flash of relief in his expression, or a grin. Then he would delete what he was trying to say and move on."

The inner circle with whom he had worked a long time, such as Thomas Hertog and Jim Hartle, were all experts at reading his face: if Stephen turned his eyes, looked at you, and slightly moved his mouth, Hertog explained, it meant he disagreed; if he lifted his eyebrows, you were on the same wavelength. However, even before she had learned how to detect more subtle expressions, Taylor discovered it was more productive to butt in and get things wrong: at least it speeded up their conversation.

Hawking once posed her a very hard problem – to find exact rotating black hole solutions of Einstein's equations – and was stunned when she came back a few days later with the solution. "I will never forget his enormous smile."

Face Reading

A vivid sense of how Stephen Hawking used more than his voice to communicate can be glimpsed in this extract of a poem, "Hawking Rises", written in 2013 by his then personal assistant, Judith Croasdell:

When he smiles

A huge beam radiates
 but then his expression quickly settles
 his eyes following the words he needs
 to speak
 he winks and beeps
 with fierce concentration

When he disapproves

His left lip goes downwards
 a grimace between disdain, distress
 and anger
 It takes you by surprise for it is terrible!
 Like a howl of rage
 even though silent

"When I started with Stephen in ninety-five", Taylor went on to explain, "that was a really important year for string theory" (which says that all matter consists of tiny, vibrating, "string-like" entities), "because there was real progress on the quantum nature of black holes using string theory. And then there were these connections between different string theories, called dualities, thanks to the work of Ed Witten" (of the Institute for Advanced Study, Princeton). Hawking's work on black hole radiation was incomplete, and string theory seemed to offer a possible new way to make some progress, and now, recalled Taylor, Hawking "expected me to jump straight to the frontier of string theory as a starting graduate student." In particular, Hawking was interested in a conjecture about M-theory that Witten had set out in 1995 while he was attending a conference at the University of Southern California, to unify all consistent versions of superstring theory. Hawking would even go on to choose the title of Taylor's thesis,

OPPOSITE Fay Dowker with Stephen Hawking in Cambridge.

Doing physics could also mean going to the pub, or eating at one of his favourite Cambridge restaurants.

ABOVE Hawking's smile, described by his former personal assistant Judith Croasdell as a "huge beam".

"Problems in M-theory". "If you look at Stephen's career over forty years, he often identified directions he wanted to learn about and then put his students to work on them."

In the six years she worked in Cambridge, Marika Taylor observed how Hawking's stubborn personality not only helped him cope with medical issues, but also made him relentless in tackling seemingly unsolvable problems. "Never giving up is the main thing Hawking has taught me – to keep attacking problems from different directions, to reach for the hardest problems and find a way to solve them."

Hawking almost treated his PhD students and collaborators as a second family, reflected Taylor.

Doing physics could also mean going to the pub, or eating at one of his favourite Cambridge restaurants. "He always made time for us, often making dignitaries wait outside his office while he talked physics with a student. He would eat lunch with us several times per week, and funded a weekly lunch for the wider group to bring everyone together."

The limitations of his speech synthesizer meant he was adept at coining pithy one-liners. "For instance, when changing his mind about what happens to information in a black hole, he announced it in the pub by turning the volume up on his synthesizer, saying simply: 'I'm coming out.'"

By then, routine encounters akin to the Nuffield workshops pioneered in Cambridge in 1981 and

Hawking's stubborn personality not only helped him cope with medical issues, but also made him relentless in tackling seemingly unsolvable problems.

1982, which had seen the most talented people from around the world meet, give lectures and work intensively on a problem, were taking place in Aspen, Colorado, or the Galileo Institute in Florence. California was another destination, with Hawking taking his students every January and February to work, bask in the sun ("it would lift his spirits") and be close to his old friends, John Preskill and Kip Thorne at Caltech, and Jim Hartle in Santa Barbara, according to Thorne, his most important collaborator.

As his daughter Lucy recalled, life in warm, dry and sunny California was "much easier for a person in a wheelchair at the time than Cambridge". Indeed, Thorne recounted how in 1991 they had even tried to lure Stephen to Caltech with a professorship. "After a considerable struggle he declined, but then we reached an agreement that instead he would come for up to a couple of months every year. He enjoyed his time out here very much."

STEPHEN HAWKING'S LEGACY

The dynasties of physicists to have emerged from Stephen Hawking's research are huge and influential. Next to the Science Museum is Imperial College, London, a scientific powerhouse on whose "Court" I am honoured to sit. When he received an honorary doctorate there, Stephen Hawking remarked, "I am proud to say that five of my greatest students have risen to become Imperial professors."

All five students are professors in the Theoretical Physics Group at Imperial College: aside from Fay Dowker, who described him in her eulogy at his funeral as "my teacher, mentor and friend", there were Jerome Gauntlett, Jonathan Halliwell, Chris Hull and Toby Wiseman.

Jerome Gauntlett became the scientific consultant on the 2014 Hawking biopic, *The Theory of Everything*, which features a scene where Hawking gets a sweater trapped over his head and has an insight that would pave the way for Hawking radiation, "widely considered to be the single most important insight into quantum gravity that has been discovered so far", said Gauntlett, who also helped to bring the star, Eddie Redmayne, up to speed with Hawking's science.

"It was a privilege to study in his research group in the exciting days when his ideas were revolutionizing our ideas of space and time and quantum theory," said Professor Chris Hull. Professor Jonathan Halliwell fondly remembered "the appealing schoolboy humour that lightened many a dull day".

Professor Toby Wiseman first encountered Stephen Hawking as a schoolboy. "I was captivated by the audacity

of the questions he was considering – how and why the universe began!" he said. "I still marvel at the remarkable insights he had."

SEE ALSO:
———————

Hawking Radiation, p.46
Caffeinated Science, p.98

ABOVE Doctoral thesis by Hawking's student Paul Shellard, entitled: *Quantum Effects in the Early Universe.*

Epilogue

While Hawking was not a fan of religion, that did not stop the interment of his ashes at Westminster Abbey in the heart of London, where some of Britain's greatest citizens are buried, such as Isaac Newton, Charles Darwin and Charles Dickens. And while he also had mixed feelings about beaming a message out into space for aliens to pick up, his voice was sent towards a nearby black hole in June 2018 during the service of thanksgiving in the Abbey, attended by his friends and colleagues, me included.

Alongside us were political leaders, celebrities such as the comedians David Walliams and Ben Miller, the model Lily Cole, the TV physicist Brian Cox, the musician Nile Rodgers, TV personality Carol Vorderman, astronaut Tim Peake, and Nick Mason and David Gilmour from Pink Floyd, along with 1,000 of 25,000 people from 100 countries who had entered a ballot to attend.

QUEEN MAB

We heard an address by Hawking's old friend, the Astronomer Royal Lord Rees, who told the congregation that "Stephen described his own scientific quest as learning the mind of God, but this was a metaphor. He shared Darwin's agnosticism, but it is fitting that he too should be interred in this national shrine." There were readings by his daughter Lucy, the actor Benedict Cumberbatch, who played the physicist in a BBC drama, and the astronaut Tim Peake, who recited "Queen Mab" by Percy Bysshe Shelley:

Earth's distant orb appeared
The smallest light that twinkles in the heaven.

Tributes came from his friend, the Nobel Prize winner Professor Kip Thorne, and Tom Nabarro, an accessibility engineer who had been left paralysed by a snowboarding accident. The ceremony began with an organ rendition of "Venus, the Bringer of Peace" from Holst's *The Planets* and closed with Wagner's "Ride of the Valkyries".

Hawking's ashes were interred in Scientists' Corner, near the graves of Sir Isaac Newton and Charles Darwin. As his daughter Lucy remarked, if her father had realized that he was lying alongside these giants of science, "he'd be overwhelmed with joy and emotion".

HERE LIES WHAT WAS MORTAL OF STEPHEN HAWKING 1942–2018, reads a Caithness slate stone depicting a black hole as a swirl of rings surrounding a darker central ellipse, along with his most famous equation, for Hawking radiation.

PARTING MESSAGE

At the memorial service, we had listened to Stephen Hawking's final message, set to an original score by the composer Vangelis, famous for his theme tunes to films like *Blade Runner* and *Chariots of Fire*, in which he talked of the preciousness of time, the challenges of climate change and the Earth's burgeoning population. "We are all time travellers journeying together into the future," he told us. "But let us work together to make that future a place we want to visit. Be brave, be determined, overcome the odds. It can be done."

That parting message was beamed into space from the European Space Agency's deep space antenna at Cebreros in Spain, towards what was then thought to be the nearest black hole, 1A 0620-00 (a binary star system, consisting of a star and a second object that cannot be seen but, from its mass, is deduced to be a black hole).

The broadcast will travel the 3,457-light-year distance and arrive in the year 5475. Stephen Hawking never himself managed to encounter one of these extraordinary and superdense objects, but one day in the far future a faint signal carrying his words will reach a black hole, the mysterious subject of his life's science.

OPPOSITE Roger Highfield's order of service for Stephen Hawking's thanksgiving ceremony.

Westminster Abbey

A Service of Thanksgiving for

PROFESSOR STEPHEN HAWKING

CH CBE FRS FRSA

8th January 1942–14th March 2018

$$T = \frac{\hbar c^3}{8\pi GMk}$$

HAWKING RADIATION

SINGULARITY

EVENT HORIZON

BLACK HOLE

COLLAPSING STAR

Friday 15th June 2018

noon

Hawking's Equation

Stephen Hawking wanted his life in science to be represented by this equation on his tombstone.

The mathematical expression shows the temperature of so-called "Hawking radiation" given off by a (non-rotating) black hole. It is a surprising yet beautiful relationship that connects thinking about the very small (quantum mechanics) with the very big (relativity), along with the science of heat and work (thermodynamics). As Hawking himself remarked in his memoir, the formula "reveals that there is a deep and previously unsuspected relationship between gravity and thermodynamics".

The ten characters in Hawking's equation express his idea that black holes in the universe are not entirely black but emit a glow that would become known as Hawking radiation.

Hawking had made it known during his sixtieth birthday celebrations that he wanted this elegant equation to adorn his tomb, perhaps because he was an admirer of the quantum pioneer Paul Dirac, whose own memorial stone in Westminster Abbey is adorned with his beautiful wave equation that unified relativity and quantum mechanics, and paved the way to discovering antimatter.

However, Hawking's wish also harked back to the great Austrian thermodynamicist Ludwig Boltzmann (1844–1906), who had an equation that defined the quantity of entropy engraved on his tomb in Vienna's Central Cemetery. Ominously, Boltzmann had subsequently hanged himself, such had been the attacks on his work during his lifetime, even though now it is accepted to have established some of the fundamentals of science. But when I pointed this out around the time of his birthday celebrations, Stephen Hawking laughed off this gloomy connection, admitting he had not known about it when conceiving the idea of his own inscription.

The ten characters in Hawking's equation express his idea that black holes in the universe are not entirely black but emit a glow that would become known as Hawking radiation.

$$T = \frac{\hbar c^3}{8\pi GMk}$$

The T stands for temperature; the $\hbar$ for Planck's constant, which is used in quantum mechanics; c for the speed of light, used in relativity theory; 8Pi helps us to grasp the black hole's spherical nature; G is Newton's constant to understand gravity; M stands for the mass of the black hole, and k for Boltzmann's constant, used in thermodynamics to relate energy to temperature for individual particles.

What this all adds up to is that the temperature is inversely proportional to the black hole's mass, which in turn means that tiny black holes are predicted to be larger emitters of Hawking radiation than larger black holes.

Brian Cox, Royal Society Professor for Public Engagement in Science, tweeted about the equation the day after the memorial service: "A wonderful moment yesterday: sitting in the Abbey debating with members of the congregation whether it would have been marginally better for the equation on Stephen's memorial stone to be S = A k c^3 / 4G(hbar) rather than T = (hbar)c^3/8(pi)GMk. And that's why I love physics."

SEE ALSO:

Hawking Radiation, p.46
On the Shoulders of Lucasian Giants, p.62
Steaming Inspiration, p.188

OPPOSITE Memorial stone to Stephen Hawking in Westminster Abbey.

Further Reading

The following is intended to point the reader in the direction of further related books and writings on Stephen Hawking and his work. It is not intended to be a comprehensive bibliography and it does not include the many specialized scientific volumes or conference proceedings that he edited, or his own scientific publications which numbered 357 between 1965 and 2018.

Books by Stephen Hawking

Hawking, Stephen. *A Brief History of Time: From Big Bang to Black Holes.* First Edition by Bantam Books, 1988.

Hawking, Stephen. *Black Holes and Baby Universes, and other Essays.* Bantam Books, 1994

Hawking, Stephen. *Stephen Hawking's Life Works: The Cambridge Lectures.* Audio Literature, 1994

Hawking, Stephen. *A Brief History of Time: From Big Bang to Black Holes. The Updated and Expanded 10th Anniversary Edition.* Bantam Books, 1996.

Hawking, Stephen. *The Universe in a Nutshell.* Bantam Books, 2001

Hawking, Stephen. *The Theory of Everything: The Origin and Fate of the Universe.* New Millennium Press, 2002

Hawking, Stephen. *A Brief History of Time: From Big Bang to Black Holes.* Updated Edition, 2011.

Hawking, Stephen. *My Brief History.* Bantam Books, 2013

Hawking, Stephen, *Black Holes – The BBC Reith Lectures*, 2016

Hawking, Stephen. *Brief Answers to the Big Questions.* Bantam Books, 2018

With Lucy Hawking

Hawking, Stephen, and Lucy Hawking. *George's Secret Key to the Universe.* Simon Schuster Books for Young Readers, 2007

Hawking, Stephen, and Lucy Hawking. *George's Cosmic Treasure Hunt.* Doubleday Children's Books, 2009

Hawking, Stephen, and Lucy Hawking. *George and the Big Bang.* Doubleday Children's Books, 2011

Hawking, Stephen, and Lucy Hawking. *George and the Unbreakable Code.* Doubleday Children's Books, 2014

Hawking, Stephen, and Lucy Hawking. *George and the Blue Moon.* Doubleday Children's Books, 2016

Hawking, Stephen, and Lucy Hawking. *George and the Ship of Time.* Simon & Schuster Books for Young Readers, 2019

Hawking, Stephen, and Lucy Hawking. *Unlocking the Universe.* Puffin. 2021

With other authors:

The Large Scale Structure of Space-Time (with G.F.R. Ellis). Cambridge 1973

Penrose, Roger, and Stephen Hawking. *The Nature of Space and Time.* Princeton University Press, 1996

Penrose, Roger. *The Large, the Small and the Human Mind.* Cambridge University Press, 1997

The Future of Spacetime (with Kip Thorne, Igor Novikov, Timothy Ferris, and Alan Lightman), 2002.

A Briefer History of Time (with Leonard Mlodinow), 2008.

The Grand Design (with Leonard Mlodinow), 2011.

Edited Compilations:

Hawking, Stephen. *On the Shoulders of Giants: The Great Works of Physics and Astronomy.* Running Press, 2002.

Hawking, Stephen. *God Created the Integers.* Running Press, 2005

Hawking, Stephen. *The Dreams That Stuff Is Made of: The Most Astounding Papers of Quantum Physics and How They Shook the Scientific World.* Running Press, 2011

Other works

Hawking, Jane. *Music to Move the Stars: A Life with Stephen.* Macmillan, 1999

Hawking: Can You Hear Me? Film. Directed by Oliver Twinch. 2021. Atlantic Productions

Hertog, Thomas. *On the Origin of Time: Stephen Hawking's Final Theory.* Penguin, 2023

Mialet, Hélène. *Stephen Hawking and the Anthropology of the Knowing Subject.* University of Chicago Press, 2012.

Mlodinow, Leonard. *Stephen Hawking: Friendship and Physics.* Penguin, 2020

About the Author

Roger Highfield is science director at the Science Museum Group, a member of the Medical Research Council, and visiting professor at University College London and the Dunn School, University of Oxford. Winner of several journalism awards in his previous roles as editor of *New Scientist* and science editor of the *Daily Telegraph*, Roger is the author of *The Physics of Christmas* and *The Science of Harry Potter*, and the coauthor of such highly acclaimed books as *The Arrow of Time, The Private Lives of Albert Einstein, Supercooperators* and *Virtual You*.

Acknowledgements

For the opportunity to write this book, thanks to the Science Museum Group, under Director Sir Ian Blatchford, notably Group Publishing Manager, Wendy Burford, Head of Cultural and Commercial Partnerships, Maren Krumdieck, and our agent, Luigi Bonomi. For his advice, key insights and generous help, thank you to our Curator of Physics, Juan-Andres Leon. For organizing my time and preserving my sanity, I am grateful to my assistants Laura Ambrose and Natasha Singh. My colleagues Juan-Andres Leon and Jenny Lawson, with help from Laura, carried out the first wave of edits and I am also grateful for the input of several other colleagues: Tom Bevan, Peter Dickinson, Anna Ferrari, Tim Leng, Andrew McLean and Dave Patten.

For his wonderful foreword, and help over the years with everything from the arrow of time to black holes and Escher, I am hugely grateful to Sir Roger Penrose, and also to his assistant Helen McGregor. For their invaluable help with understanding his research, and how it was done, many thanks to Stephen Hawking's former students, Fay Dowker, Gary Gibbons, Thomas Hertog (thanks also to Thomas for an advance copy of his *On the Origin of Time. Stephen Hawking's Final Theory*) Paul Shellard and Marika Taylor, along with his old friend Lord (Martin) Rees. For helping me to get the science right, thanks again to his former students and also to Sam Braunstein, Jim Hartle, Alexey Miekhin, Neil Turok, Michael S. Turner and Kip Thorne. For revealing the complex history of his synthesized voice, I am grateful to Eric Dorsey, Jonathan Wood and Walter Woltosz. For their help identifying synthesizers and communication equipment, Mark Green, Robert Maskell and Lama Nachman from Intel Corporation.

For insights into the contents and organization of the office, thank you to his former assistant, Judith Croasdell. For his roles in *The Simpsons* and *The Big Bang Theory*, it was an honour and pleasure to deal with Al Jean and Bill Prady, respectively (and with Sean Carroll). Thanks also to former Science Museum employee, now Hawking's official biographer, Graham Farmelo, for being so generous with his time and advice. For explaining how he got run over by Stephen Hawking in the name of charity, I am grateful to Brian Cox. And for recounting how their paths harmoniously crossed Stephen Hawking's orbit, thank you to Rena Harms, Dina and Prakash Nayee, and Layla Sarakalo.

For providing a historical perspective, I owe a debt of gratitude to Jürgen Renn and John Heilbron. For providing an artistic perspective, thank you to Paul Gopal-Chowdhury. Many of the above also critiqued draft versions, though any outstanding errors are, of course, entirely my own. For allowing me to quote from his biopic, "Hawking: Can you hear me?", thank you to Anthony Geffen of Atlantic. For all their hard work in helping to lay the foundations for this book, I would also like to thank the Science Museum Group conservation team: Beth Baker, Laura Chaillie, Jessica Crann, Sophie Croft, Lorna Flynn, Marisa Kalvins, Liza Nathan, Ruth Nightingale, Kirsten Strachan, Emily Yates and Matt Walker; our photographers, Isidora Bojovic, Jennie Hills, Nicola Onions, Jamie Torrance and Andrew Tunnard; and our communications team, Chloë Abley, William Dave and Peter Dickinson. When it comes to documentation, a particular thanks to Rowena Hartley; for registration Nicole Simoes da Silva; and for logistics, Halina Bartoszewska, Shona Holden and Laura Waters. For facilitating the visits of experts to view Hawking objects at the Science Museum Group's National Collections Centre, Alison Faraday and Laura Humphreys. For the Stephen Hawking at Work exhibition touring all our Science Museum Group sites: Rachel Bateson, Imogen Clarke, David Dewhurst, Nicky Lacourse, Elisenda Losantos, Lee Roberts and Lorraine Ward.

For all their wise guidance, thank you to our Head of Collections, Jessica Bradford, and Jane Desborough, Keeper of Science Collections and chair of the Hawking project board. I would also especially like to thank my former colleagues, Tilly Blyth and Ali Boyle, for their monumental work on acquiring the office in the first place, and their brilliant insight identifying it as an object of cultural significance in its own right. For their roles in producing this book, I am hugely grateful to my colleagues at Dorling Kindersley, Pete Jorgensen along with Jo Connor, Graham Coster, Anna Formanek, Eoghan O'Brien, Sakshi Saluja, Rituraj Singh and Flo Ward.

Thank you to the Hawking family, notably Lucy Hawking. Finally, a heartfelt thank you to my family – Julia, Holly and Rory – for putting up with me during my tenth attempt to write a popular science book.

Index

Page references in *italics* indicate illustrations.

Picture Credits

The publisher would like to thank the following for their kind permission to reproduce their photographs:

(Key: a-above; b-below/bottom; c-centre; f-far; l-left; r-right; t-top)

2 Alamy Stock Photo: Danita Delimont / Alison Wright. **7 Alamy Stock Photo:** Andrew Fox (cla). **22-23 ScanLAB Projects. 24-25 ScanLAB Projects. 26 Alamy Stock Photo:** ARCHIVIO GBB (cla). **27 Alamy Stock Photo:** Krzysztof Jakubczyk (br). © **Science Museum Group:** Fred Cuming RA (tl). **The Vatican Pontifical Academy of Sciences:** (crb). **28** © **Science Museum Group:** Copyright of St Albans School. **29 Alamy Stock Photo:** ARCHIVIO GBB (br). **31 Alamy Stock Photo:** Krzysztof Jakubczyk (t). **33 Alamy Stock Photo:** Krzysztof Jakubczyk (cb). **36 Getty Images:** Archive Photos / Harvey Meston (tl). **37 Science Photo Library:** Max-planck-institut Fur Extraterrestrische Physik (bl). **39 John Cairns & permission of Oxford Mathematics, 2017:** (br). **40** © **Science Museum Group:** IOP Publishing / NSF / LIGO / Sonoma State University / A. Simonnet. **42-43 Science Photo Library:** Caltech / MIT / LIGO Lab. **48 AIP Emilio Segr Visual Archives, Physics Today Collection:** (tr). **49 Science Photo Library:** European Southern Observatory (t). **51 Alamy Stock Photo:** Heritage Image Partnership Ltd / © Fine Art Images (br). **54 The Vatican Pontifical Academy of Sciences. 58** © **Science Museum Group:** Cambridge University Press / Reproduced with permission of the Licensor through PLSclear / The Very Early Universe. **63** © **Science Museum Group:** © 1962 by The Regents of the University of and California, published by the University of California Press,. **64 Getty Images:** Bettmann (b); Hulton Fine Art Collection / brandstaetter images (tr). **67** © **Science Museum Group:** Designed by Glazier Design / Isle of Man Post Office. **70 Alamy Stock Photo:** Alpha Historica (tl). **75** © **Science Museum Group:** Fred Cuming RA. **77** © **Science Museum Group:** © David Hockney. **78-79** © **Science Museum Group:** Tan Swie Hian. **83 Science Photo Library:** CERN. **85 Alamy Stock Photo:** Pictorial Press Ltd (br). **87 Getty Images:** Kimberly White (t). **95 Getty Images:** Tim P. Whitby (br). **99 Alamy Stock Photo:** Liam White (tr). **108 Matt Groening The Simpsons and © 2024 20th Television:** (t). **112 Alamy Stock Photo:** Horst Friedrichs (tl). **Getty Images:** CBS Photo Archive / Sonja Flemming (crb). **113** © **Science Museum Group:** Peter Dean (tl); NASA / JPL / APL / SwRI (crb). **117 Alamy Stock Photo:** Xinhua (tr). **118** © **Science Museum Group:** Photographed by Milton H. Greene · Compilation © 2023 Joshua Greene · www.miltonhgreene.com (b). **121** © **Science Museum Group:** Welbeck Publishing Group Ltd. **125 Alamy Stock Photo:** Pictorial Press Ltd. **130 Getty Images:** Menahem Kahana. **131 Alamy Stock Photo:** Science History Images / Photo Researchers (tl). **134 Scott Noble / NASA-GSFC:** (cra). **135 Alamy Stock Photo:** PA Images / Haydn West (tl). **141** © **Science Museum Group:** Jaime Travezan. **143** © **Science Museum Group:** Matt Groening The Simpsons and © 2024 20th Television. **144** © **Science Museum Group:** Matt Groening The Simpsons and © 2024 20th Television (bl, br). **145** © **Science Museum Group:** Matt Groening The Simpsons and © 2024 20th Television. **149** © **2023 CERN:** ATLAS Experiment (tr). **150** © **Science Museum Group:** NASA. **154 Science Photo Library:** WMAP Science Team, NASA (t). **163 Alamy Stock Photo:** Horst Friedrichs (b). **Khristopher Kabbabe:** (tr). **164 Alamy Stock Photo:** NASA Image Collection (b). **166 Alamy Stock Photo:** NG Images (t). © **Science Museum Group:** Virgin Galactic (bl). **173** © **Science Museum Group:** Peter Dean. **181 Getty Images:** CBS Photo Archive / Sonja Flemming. **183** © **Science Museum Group:** Design © Mark Champkins. **185** © **Science Museum Group:** Maureen Short (t). **187** © **Science Museum Group:** NASA / JPL / APL / SwRI. **188** © **Science Museum Group:** Corgi (b). **194** © **Science Museum Group:** Universal Music Group. **199 Getty Images:** Sion Touhig (tl). © **Science Museum Group:** Pride of Britain Awards 2023 (br). **201 Alamy Stock Photo:** Miguel Sayago (t). **211 Alamy Stock Photo:** Liam White (tr). **213** © **Science Museum Group:** Pride of Britain Awards 2023. **218 Don Page:** (tr). **219 Alamy Stock Photo:** Mirrorpix / Trinity Mirror. **221 Getty Images:** David Montgomery. **223 Anna Zytkow. 224 Getty Images:** Sion Touhig (t). **229 Getty Images:** Peter Dazeley. **232** © **Science Museum Group:** (cla)

Cover images: *Front:* **Getty Images:** CBS Photo Archive / Sonja Flemming cla; © **Science Museum Group:** Photographed by Milton H. Greene · Compilation © 2023 Joshua Greene · www.miltonhgreene.com crb; *Back:* **Alamy Stock Photo:** ARCHIVIO GBB tl, Krzysztof Jakubczyk bc; **Getty Images:** Tim P. Whitby cra; © **Science Museum Group:** Peter Dean tr, Fred Cuming RA tc, NASA clb

All other images © Dorling Kindersley

Acknowledgement of the Hawking family

In 2021 the Science Museum Group acquired Stephen Hawking's Cambridge University office for the nation. The entire contents are available on the Group's Collections Online website, inspiring people with Hawking's intellect, curiosity, determination and achievements.

The Science Museum Group would like to acknowledge the enthusiasm and generosity of Lucy, Robert and Tim Hawking, who enabled the acquisition of this collection via the Acceptance in Lieu scheme. The Group thanks the Hawking family for their continued input and valued guidance in the development of the collection.

Object no.2021-561: Stephen Hawking's Office
Accepted in lieu of inheritance Tax by HM Government from the Estate of Stephen Hawking and allocated to the Science Museum, 2021.

Editor Florence Ward
Senior Art Editor Anna Formanek
Designer Eoghan O'Brien
Senior Production Editor Marc Staples
Senior Production Controller Louise Minihane
Managing Editor Pete Jorgensen
Managing Art Editor Jo Connor
Publisher Mark Searle

DK would like to thank Caroline West for proofreading
and Emma Caddy for indexing.

First published in Great Britain in 2024 by
Dorling Kindersley Limited
One Embassy Gardens, 8 Viaduct Gardens,
London SW11 7BW

The authorised representative in the EEA is
Dorling Kindersley Verlag GmbH. Arnulfstr. 124,
80636 Munich, Germany

Page design copyright © 2024 Dorling Kindersley Limited
A Penguin Random House Company

Texts © SCMG Enterprises Ltd, 2023
Science Museum ® SCMG Enterprises Ltd and designs
© SCMG Enterprises Ltd

10 9 8 7 6 5 4 3 2 1
001–336462–Mar/2024

A CIP catalogue record for this book
is available from the British Library.
ISBN: 978-0-2416-3066-2

Printed and bound in China

www.dk.com

In association with
The Science Museum Group
Exhibition Road
London SW7 2DD
Science Museum logo © 2024 Science Museum
www.sciencemuseum.org.uk

Every purchase supports the museum.

This book was made with Forest
Stewardship Council™ certified
paper – one small step in DK's
commitment to a sustainable future.
**For more information go to
www.dk.com/our-green-pledge**

SYMMETRY

tein

tp. Squid ?

have tea

the kentving

chancellor
of the
exchequer

KNOW
ABOUT
AMA

INSEC
TONY

I AM

you get
an
else

ROYAL
DEAD

n?

D N

TAB C

TOWNSEND

DON'T
CER

WEST

Ra

JOHN HD

CALIFORNIA INSTITUTE

MASSACHUSETTS HOME FOR

S AT ANY TIME

HRE RES

WORK YOUR

HENRY
WA HERE

HOPE

THE FIRE
AN ARROW
FLIES

ONE OF THE
FELLS

Z UMON

L
IT IS
THE
END

HAPPY

CALCULATOR
AS A CHECK

CZECH AS A
CALCULATOR